100 MATHS HOMEWORK ACTIVITIES

D0545763

CONTENTS

Published by
Scholastic Ltd,
Villiers House,
Clarendon Avenue,
Leamington Spa,
Warwickshire CV32 5PR

© Scholastic Ltd 2001
Text © Sue Gardner and Ian Gardner 2001
Additional material on
pages 6–8 © Ann Montague-Smith 2001

   5 6 7 8 9   4 5 6 7 8 9 0

AUTHORS
Sue Gardner and Ian Gardner

EDITORIAL & DESIGN
Crystal Presentations Ltd

COVER DESIGN
Joy Monkhouse

ILLUSTRATOR
Theresa Tibbetts

Acknowledgements

The publishers wish to thank:
Ann Montague-Smith for her invaluable advice in
the development of this series.
The Controller of HMSO and the DfEE for the use
of extracts from *The National Numeracy Strategy:
Framework for Teaching Mathematics* © March 1999,
Crown Copyright (1999, DfEE, Her Majesty's
Stationery Office).

British Library Cataloguing-in-Publication Data

A catalogue record of this book is available from the British
Library.

ISBN 0-439-01846-3

The right of Sue Gardner and Ian Gardner to be
identified as the Authors of this work has been asserted
by them in accordance with the Copyright, Designs and
Patents Act 1988.

PAGE IN THIS BOOK	100 MATHS HOMEWORK ACTIVITIES YEAR 3		NATIONAL NUMERACY STRATEGY		100 MATHS LESSONS		
	ACTIVITY NAME	HOMEWORK	STRAND	TOPIC	NNS UNIT	LESSON	PAGE
29	Order by phone	Practice exercise	Numbers and the number system	Place value and ordering	①	①	⑲ ⑳
30	Numbers on a plate	Practice exercise	Numbers and the number system	Place value and ordering	①	②	⑳ ㉑
31	Rounding game	Maths to share	Numbers and the number system	Estimating and rounding	①	③	㉑
32	It all adds up	Maths to share	Calculations	Mental calculation strategies (+ and −)	② ③	①	㉕ ㉖
33	Target 20	Maths to share	Calculations	Rapid recall of addition and subtraction facts	② ③	②	㉖ ㉗
34	Make it pay (1)	Maths to share	Solving problems	Problems involving money	② ③	④	㉗ ㉘
35	Number sentences	Practice exercise	Calculations	Understanding addition and subtraction	② ③	⑤	㉘ ㉙
36	Make that total	Practice exercise	Calculations	Understanding addition and subtraction	② ③	⑦	㉙ ㉚
37	Make 20	Maths to share	Calculations	Rapid recall of addition and subtraction facts	② ③	⑧	㉚ ㉛
38	What's in a name?	Practice exercise	Measures, shape and space	Shape and space	④ ⑥	①	㉞ ㉟
39	All in shape (1)	Maths to share	Measures, shape and space	Shape and space	④ ⑥	⑤	㊱ ㊲
40	All in shape (2)	Maths to share	Measures, shape and space	Shape and space	④ ⑥	⑥	㊲ ㊳
41	Look at the label	Maths to share	Measures, shape and space	Measures	④ ⑥	⑨	㊴ ㊵
42	How long?	Maths to share	Measures, shape and space	Measures	④ ⑥	⑪	㊵ ㊶
43	More measures	Maths to share	Measures, shape and space	Measures	④ ⑥	⑫	㊵ ㊶
44	Longer – shorter	Maths to share	Measures, shape and space	Measures	④ ⑥	⑬	㊵ ㊶
45	How many?	Maths to share	Numbers and the number system	Counting, properties of numbers and number sequences	⑧	①	㊾ ㊿
46	Sequences (1)	Practice exercise	Numbers and the number system	Counting, properties of numbers and number sequences	⑧	②	㊿ 51
47	Pigs and ducks	Games and puzzles	Solving problems	Reasoning about numbers	⑧	④	51 52
48	Target 80	Games and puzzles	Solving problems	Reasoning about numbers	⑧	⑤	52 53
49	Superbugs!	Practice exercise	Calculations	Understanding multiplication and division	⑨ ⑩	①	56
50	Supershapes!	Practice exercise	Calculations	Understanding multiplication and division	⑨ ⑩	②	57
51	Fair shares	Maths to share	Calculations	Understanding multiplication and division	⑨ ⑩	④	58 59
52	Making money	Practice exercise	Solving problems	Problems involving money	⑨ ⑩	⑧	60 61
53	What shall I buy?	Games and puzzles	Solving problems	Problems involving money	⑨ ⑩	⑩	61

PAGE IN THIS BOOK	ACTIVITY NAME	HOMEWORK	STRAND	TOPIC	NNS UNIT	LESSON	PAGE
54	Shape share	Practice exercise	Numbers and the number system	Fractions	11	1	66 67
55	Dividing out	Practice exercise	Numbers and the number system	Fractions	11	3	67 68
56	Part shares	Practice exercise	Numbers and the number system	Fractions	11	5	69
57	Jumps	Practice exercise	Calculations	Pencil and paper procedures (+ and –)	12	1	71 72
58	Giant jumps	Practice exercise	Calculations	Pencil and paper procedures (+ and –)	12	2	72 73
59	About time	Maths to share	Measures, shape and space	Measures	12	3	73 74
60	TV (1)	Maths to share	Handling data	Organising data	13	1	75 76
61	TV (2)	Maths to share	Handling data	Organising data	13	2	76 77
62	TV (3)	Maths to share	Handling data	Organising data	13	3	77 78
63	Going up	Practice exercise	Numbers and the number system	Place value and ordering	1	2	88 89
64	Scales	Maths to share	Measures, shape and space	Measures	1	3	89
65	Make it up	Practice exercise	Calculations	Understanding addition and subtraction	2 3	1	92
66	Number pairs	Practice exercise	Calculations	Understanding addition and subtraction	2 3	2	92 93
67	Make it pay (2)	Practice exercise	Solving problems	Problems involving money	2 3	4	93 94
68	Making more money	Practice exercise	Solving problems	Problems involving money	2 3	6	95 96
69	One stop party shop!	Maths to share	Solving problems	Problems involving money	2 3	8	96 97
70	Word problems!	Practice exercise	Solving problems	Problems involving 'real life'	2 3	9	97
71	Shapes in words	Practice exercise	Measures, shape and space	Shape and space	4 6	1	103 104
72	Symmetry	Maths to share	Measures, shape and space	Shape and space	4 6	3	104 105
73	About time	Maths to share	Measures, shape and space	Measures	4 6	5	106 107
74	Picture this	Practice exercise	Solving problems	Reasoning about shapes	4 6	6	107 108
75	Boxes	Games and puzzles	Measures, shape and space	Shape and space	4 6	9	108 109
76	On the grid	Practice exercise	Measures, shape and space	Shape and space	4 6	10	109
77	Getting into shape	Practice exercise	Measures, shape and space	Shape and space	4 6	13	111
78	Sequences (2)	Practice exercise	Numbers and the number system	Counting, properties of numbers and number sequences	8	1	119 120

100 MATHS HOMEWORK
ACTIVITIES YEAR 3

NATIONAL NUMERACY STRATEGY

100 MATHS LESSONS

REFERENCE GRID

REFERENCE GRID

PAGE IN THIS BOOK	ACTIVITY NAME	HOMEWORK	STRAND	TOPIC	NNS UNIT	LESSON	PAGE
79	Sequences (3)	Practice exercise	Numbers and the number system	Counting, properties of numbers and number sequences	8	4	121 122
80	Statements	Practice exercise	Calculations	Understanding multiplication and division	9 10	1	125 126
81	Take back	Practice exercise	Calculations	Understanding multiplication and division	9 10	3	126 127
82	Differences	Games and puzzles	Calculations	Understanding addition and subtraction	9 10	4	127
83	Money problems!	Practice exercise	Solving problems	Problems involving money	9 10	5	127 128
84	This way, that way	Practice exercise	Calculations	Understanding addition and subtraction	9 10	7	129
85	Number machines	Practice exercise	Calculations	Mental calculation strategies (+ and –)	9 10	8	129 130
86	Coin combos	Practice exercise	Solving problems	Problems involving money	9 10	10	131
87	Fraction machine	Practice exercise	Numbers and the number system	Fractions	11	3	135 136
88	Shape split	Practice exercise	Numbers and the number system	Fractions	11	4	136 137
89	All in a day	Investigation	Handling data	Organising data	12	3	140 141
90	Birthday chart	Investigation	Handling data	Organising data	12	4	141
91	Pets corner	Investigation	Handling data	Organising data	12	5	142
92	House!	Games and puzzles	Numbers and the number system	Estimation and rounding	1	1	149 150
93	Rough talk	Practice exercise	Numbers and the number system	Estimation and rounding	1	2	150 151
94	Staging posts (1)	Practice exercise	Calculations	Understanding addition and subtraction	2 3	1	153
95	Budget	Maths to share	Calculations	Pencil and paper procedures (+ and –)	2 3	2	154
96	All things equal	Practice exercise	Calculations	Understanding addition and subtraction	2 3	4	155
97	How's that? (1)	Practice exercise	Calculations	Understanding addition and subtraction	2 3	5	155 156
98	How much?	Practice exercise	Calculations	Pencil and paper procedures (+ and –)	2 3	6	156 157
99	Staging posts (2)	Practice exercise	Calculations	Pencil and paper procedures (+ and –)	2 3	7	157
100	Giving change	Maths to share	Solving problems	Problems involving money	2 3	8	157 158
101	Problems, problems!	Games and puzzles	Solving problems	Reasoning about numbers	2 3	10	159
102	Solid shapes	Practice exercise	Measures, shape and space	Shape and space	4 6	1	162
103	2-D shapes	Practice exercise	Measures, shape and space	Shape and space	4 6	2	163

	100 MATHS HOMEWORK ACTIVITIES YEAR 3		NATIONAL NUMERACY STRATEGY		100 MATHS LESSONS		
PAGE IN THIS BOOK	ACTIVITY NAME	HOMEWORK	STRAND	TOPIC	NNS UNIT	LESSON	PAGE
104	In balance	Practice exercise	Measures, shape and space	Shape and space	4 6	3	163 164
105	Same and different	Games and puzzles	Solving problems	Reasoning about shapes	4 6	4	164 165
106	All right now	Practice exercise	Measures, shape and space	Shape and space	4 6	6	165 166
107	Seeing double	Practice exercise	Measures, shape and space	Measures	4 6	7	166 167
108	Treasure Island	Practice exercise	Measures, shape and space	Shape and space	4 6	9	167 168
109	In a spin	Practice exercise	Measures, shape and space	Shape and space	4 6	11	168 169
110	Full to capacity	Practice exercise	Measures, shape and space	Measures	4 6	12	169 170
111	Number sort	Practice exercise	Numbers and the number system	Place value and ordering	8	1	177 178
112	How's that? (2)	Practice exercise	Solving problems	Reasoning about numbers	8	2	178 179
113	Odds and evens	Games and puzzles	Solving problems	Reasoning about numbers	8	3	179 180
114	Starting grid	Games and puzzles	Solving problems	Reasoning about numbers	8	5	180
115	What's the story?	Practice exercise	Solving problems	Making decisions	9 10	1	183 184
116	What's it worth?	Games and puzzles	Solving problems	Problems involving money	9 10	5	185 186
117	Maths links	Practice exercise	Calculations	Understanding multiplication and division	9 10	6	186
118	Remainders (1)	Practice exercise	Calculations	Understanding multiplication and division	9 10	7	187
119	Remainders (2)	Practice exercise	Calculations	Understanding multiplication and division	9 10	9	188 189
120	Fraction pieces (1)	Practice exercise	Numbers and the number system	Fractions	11	1	192 193
121	Which is bigger?	Practice exercise	Numbers and the number system	Fractions	11	3	193 194
122	Fraction pieces (2)	Practice exercise	Numbers and the number system	Fractions	11	4	194 195
123	Fraction match	Practice exercise	Numbers and the number system	Fractions	11	5	195
124	Number crunch	Practice exercise	Calculations	Mental calculation strategies (+ and −)	12	1	197 198
125	Take that	Practice exercise	Calculations	Pencil and paper procedures (+ and −)	12	4	199 200
126	Time diary	Investigation	Measures, shape and space	Measures	12	5	200
127	Drinks all round	Investigation	Handling data	Organising data	13	1	201 202
128	Favourites	Practice exercise	Handling data	Organising data	13	2	202 203

100 MATHS HOMEWORK ACTIVITIES

100 Maths Homework Activities is a series of teachers' resource books for Years 1–6. Each book is year-specific and provides a core of homework activities for mathematics within the guidelines for the National Numeracy Strategy in England. The content of these activities is also appropriate for and adaptable to the requirements of Primary 1–7 in Scottish schools.

Each book offers three terms of homework activities, matched to the termly planning in the National Numeracy Strategy *Framework for Teaching Mathematics* for that year. Schools in England and Wales that decide not to adopt the National Numeracy Strategy will still find the objectives, approaches and lesson contexts familiar and valuable. However, the teacher will need to choose from the activities to match their own requirements and planning.

The homework activities provided in the books are intended as a support for the teacher, school mathematics leader or trainee teacher. The series can be used alongside its companion series, *100 Maths Lessons and more*, or with any mathematics scheme of work, as the basis for planning homework activities throughout the school, in line with the school's homework policy. The resources can be used by teachers with single-age classes, mixed-age, single- and mixed-ability groups and for team planning of homework across a year or key stage. The teacher may also find the activities valuable for extension work in class or as additional resources for assessment.

Using the books

The activities in this book are for Year 3/Primary 3–4 classes and are a mix of mathematics to share with a helper – a parent, neighbour or sibling, games and puzzles to do at home, practice exercises, some 'against the clock', and activities to help the children develop mathematics investigation skills. The activities have been chosen to ensure that each strand and topic of the National Numeracy Strategy *Framework for Teaching Mathematics* is included and that the children have opportunities to develop their mental strategies, use paper-and-pencil methods appropriately, and use and apply their mathematics to solve problems.

Each of the 100 homework activities in this book includes a photocopiable page to send home. The page provides instructions for the child and a brief explanation for a helper, stating simply and clearly its purpose and suggesting support and/or a further challenge to offer the child. The mathematics strand and topic addressed by each activity and the type of homework being offered are indicated on each page. The types are shown by the following symbols:

maths to share	games and puzzles	practice exercise	investigation	timed practice exercise

There is a supporting teacher's note for each activity. These notes include:

- **Learning outcomes:** the specific learning objectives of the homework (taken from the National Numeracy Strategy *Framework for Teaching Mathematics*);
- **Lesson context:** a brief description of the classroom experience recommended for the children prior to undertaking the homework activity;
- **Setting the homework:** advice on how to explain the work to the children and set it in context before it is taken home;
- **Back at school:** suggestions for how to respond to the returned homework, such as discussion with the children or specific advice on marking, as well as answers, where relevant.

Supporting your helpers

Extensive research by the IMPACT Project (based at University of North London) has demonstrated how important parental involvement is to children's success in mathematics. A homework diary photocopiable sheet is provided on page 8 that can be sent home with the homework. This sheet has room for records of four pieces of homework and can be kept in a file or multiple copies stapled together to make a longer-term homework record. For each activity, there is space to record of the name of the activity and the date when it was sent home, and spaces for a brief comment from the helper, the child and the teacher on their responses to the work. The homework diary is intended to encourage home–school links, so that parents and carers know what is being taught in school and can make informed comments about their child's progress.

Name _____					
Name of activity & date sent home	Helper's comments	Child's comments			Teacher's comments
		Did you like this? Colour a face.	How much did you learn? Colour a face.		
ROUNDING GAME	NADINE ENJOYED THIS GAME AND WANTED TO KEEP PLAYING! SHE DIDN'T FIND THE MATHS HARD THIS TIME.	a lot ☺ / a little 😐 / not much ☹	a lot ☺ / a little 😐 / not much ☹		Excellent! Nadine has really gained in confidence lately. You could try her with some rounding of really big numbers (to the nearest 10).
IT ALL ADDS UP	NADINE FOUND LOTS OF WAYS AND SHE TRIED TO FIND THEM ALL.	a lot ☺ / a little 😐 / not much ☹	a lot ☺ / a little 😐 / not much ☹		Nice to see Nadine working in such an organised way. Thank you.

Using the activities with *100 Maths Lessons* series

The organisation of the homework activities in this book matches the planning grids within *100 Maths Lessons and more: Year 3* (also written by Sue Gardner and Ian Gardner and published by Scholastic), so that there is homework matching the learning objectives covered in each unit of work in each term. Grids including details of which lessons in *100 Maths Lessons and more: Year 3* have associated homework activities in *100 Maths Homework Activities*, with the relevant page numbers, are provided on pages 2–5 in this book to help teachers using *100 Maths Lessons and more: Year 3* with their planning.

About this book: Year 3/Primary 3–4

This book contains five types of homework activities, all recommended by the NNS for children in Y3/Primary 3–4: 'Maths to share', 'Investigations', 'Games and puzzles' and 'Practice exercises', some as timed extension exercises. All the 'Maths to share' activities encourage the child and a helper to work together to carry out the tasks within home contexts. The 'Investigations' and 'Games and puzzles' involve a mix of mental arithmetic and strategy games for two or more players. Where there are specific solutions to these activities, it is recommended that these should be sent home at a later date. The 'Practice exercises' provide children with opportunities for further practice of number and 'four rules' work done in school and to keep mental arithmetic skills sharp. These activities are designed for children to work on independently at their own pace, or 'against the clock' as an extension activity. For these activities, helpers are given guidance on ways to help if their child gets stuck.

Equipment and resources

The majority of these homework activities require little or no resources. Sometimes they specify the use of household equipment or resources that are commonly available at home. Some of the activities require the use of classroom resources such as a number grid and different sets of numeral cards.

Reading the instructions on the homework sheets

In Year 3/Primary 3–4 classes, children are at different stages of learning to read. In most cases, the language demands are kept to a realistic level. There may be occasions, however, where the parent can reasonably be expected to clarify certain words, particularly those with a precise mathematical definition. To help the child read the instructions on the homework sheets, it is suggested that some guidance is given at the time that the homework is set.

Year 3 reporting

Children in England entering Y3 will have recently had their performance in mathematics reported on the basis of national tests and their teachers' own judgements of their current abilities. For those children who perform above the average levels, these homework tasks offer further opportunities to consolidate and extend their understanding. For children who lack confidence, or have specific difficulties in mathematics, the opportunity to work on tasks at their own pace and in a comfortable environment can be particularly beneficial. Whatever the ability of the child, the emphasis of such work should be to develop children's confidence and understanding of the different aspects of mathematics, and to give opportunities for parents and children to enjoy doing different mathematical activities together.

Name _____

Name of activity & date sent home	Helper's comments	Child's comments		Teacher's comments
		Did you like this? Colour a face.	**How much did you learn?** Colour a face.	
		☺ a lot ☺ a little ☹ not much	☺ a lot ☺ a little ☹ not much	
		☺ a lot ☺ a little ☹ not much	☺ a lot ☺ a little ☹ not much	
		☺ a lot ☺ a little ☹ not much	☺ a lot ☺ a little ☹ not much	
		☺ a lot ☺ a little ☹ not much	☺ a lot ☺ a little ☹ not much	

Teachers' notes

TERM 1

p29 ORDER BY PHONE
PRACTICE EXERCISE

Learning outcomes
- **Order whole numbers to at least 100.**
- Read and begin to write the vocabulary of comparing and ordering numbers, including ordinal numbers to at least 100.

Lesson context
Present the class with three single-digit numbers. Ask the children to combine two to make two-digit numbers (e.g. 7 and 4 makes 47 or 74). Once all the possible numbers are listed, reorder these in ascending numerical order. Set groups working with their own set of three or four numbers to 'mix and match'.

Setting the homework
Ask the children to use their own telephone number, or one with which they are familiar. Discourage them from using area codes and mobile phone numbers as this creates too many potential combinations.

Back at school
Look at strategies for methodically working through the task (e.g. combining the first digit with the second, then the third, fourth, fifth and so on). This will help to reinforce this important element of using and applying mathematics. You should also confirm that the children understand the place value significance of the digits.

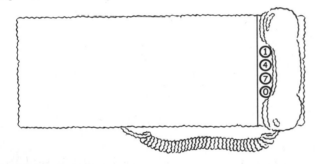

p30 NUMBERS ON A PLATE
PRACTICE EXERCISE

Learning outcomes
- **Read and write whole numbers to at least 1000** in figures and words.

Lesson context
As a class, involve everyone in using place value cards to create two- and three-digit numbers. Ask related questions: *What is the tens digit worth?* Provide some numbers for individuals/pairs to create using the place value cards. Once made these can be partitioned and recorded (e.g. 68 is 60 and 8). Additionally/alternatively, the numbers can be written in words.

Setting the homework
Ask the children to use a registration plate with which they are familiar. Discourage them from using more/fewer digits from old or personalised number plates as this will create too many or insufficient combinations.

Back at school
Pay particular attention to the spelling of irregular words such as 'forty' (not 'four-ty'). You might also consider what the maximum number of combinations is with three digits. Avoid 0 and repeated digits. For 2, 8 and 4 the six possible combinations are: 248, 284, 428, 482, 824, 842.

p31 ROUNDING GAME
MATHS TO SHARE

Learning outcome
- Round any two-digit number to the nearest 10.

Lesson context
Talk about the conventions of rounding a two-digit number up or down to the nearest multiple of 10. Use a class number line for demonstration. Clarify the convention of rounding up two-digit numbers ending in a 5. Provide some numbers up to 100 (and possibly beyond), each time asking for that number to the nearest multiple of 10. Place the children into groups of three or four and give each a standard dice. Tell them to take turns to create a two-digit number (e.g. 3 and 2 makes 32 or 23). This is rounded to the nearest 10. The winner is the first to score every multiple of 10 to 70.

Setting the homework
Explain how to play the activity detailed on the homework sheet, using a book for demonstration purposes. The children will also need to be briefed on how to complete the recording sheet, as this will be the only evidence of understanding.

Back at school
Talk about the activity. Try a few three-digit numbers which involve rounding up to the nearest 100.

p32 IT ALL ADDS UP
MATHS TO SHARE

Learning outcome
- Use knowledge that addition can be done in any order to do mental calculations more efficiently. For example, put the larger number first and count on.

Lesson context
Work with the whole class on bonds to 10. Make use of interlocking cubes in sticks of 10. Use five cubes in two colours to emphasise numbers such as 7 as '5 and a bit'. This is useful in recognising the commutative aspect of adding the numbers (i.e. order is not significant). Organise the children in pairs to practise complements of 10 or higher.

Setting the homework
Talk over the sheet, referring to the idea of methodical and systematic working. Remind the pupils of how they may have approached page 29, if it has been used recently.

Back at school
Talk about the number of different arrangements found. Identify whether there is a best way or whether, as is often the case, this is open to discussion.

p33 TARGET 20

MATHS TO SHARE

Learning outcomes
- **Know by heart all addition facts for each number to 20.**
- Use knowledge that addition can be done in any order to do mental calculations more efficiently.

Lesson context
Practise counting in 2s, then 3s. Look at the odd/even alternation of the 3 times table. Ask the children to make small totals by combining 2s and 3s. Organise the children into working groups and ask them to make given totals such as 15 or 20 using 2s and 3s in combination.

Setting the homework
Model the homework activity by using a large set of teaching numeral tiles from 0 to 9. One child selects a tile, the second child selects another tile and finds the total. The third child now tries to make up the total to 20 with the final tile. If this is not possible, they can try a fourth tile.

Back at school
Use this activity to consider how many different ways there are of making 20 with three different digit cards. Again, systematic ways of working will help here.

p34 MAKE IT PAY (1)

MATHS TO SHARE

Learning outcomes
- Recognise all coins (to 20p).
- Solve problems involving money, including finding totals and working out which coins to pay.

Lesson context
Use large representations of coins to remind the class of the various coin denominations. Ensure that the children are able to recognise all coins to at least 20p, and begin to see the inter-relationships between the different denominations. Provide groups with sets of picture cards featuring items with suitable prices attached. The task is to select any two, find the total and to record the coins used.

Setting the homework
Provide totals up to the value of 20p and, for each case, identify which coins would be used to make that total. Emphasise the importance of working with a helper, wherever this is possible.

Back at school
Consider which amounts required the most coins. The number of coins required for any total to 20p should never exceed four.

p35 NUMBER SENTENCES

PRACTICE EXERCISE

Learning outcomes
- Extend understanding that subtraction is the inverse of addition.
- Check subtraction with addition.

Lesson context
Display cards featuring 3, 7, 10 , +, − and =. Invite individuals to create number stories from this selection. Use this as an opportunity to revise the language of computation, especially subtraction. Provide groups with a suitable set of related tiles, encouraging them to look for all combinations.

Setting the homework
Provide an example based on the pentagon shape illustrated on the homework sheet. Add the following numbers (one at each corner): 13, 7, 8, 6, 5. Ask the children to create number sentences for addition and subtraction (e.g. $13 − 7 = 6$).

Back at school
If the guidance notes have been followed, the children should find eight different solutions, four for each operation of addition and subtraction.

p36 MAKE THAT TOTAL

PRACTICE EXERCISE

Learning outcomes
- Understand that more than two numbers can be added (Y2 revision).
- Use patterns of similar calculations.

Lesson context
Working with the whole class, provide some target numbers, to be created by combining, say, 3s and 5s only. A target such as 13, for example, can be generated by combining two 5s and one 3. When children are confident with this, they should begin to make totals mentally. Activities of this type could usefully be explored by groups, adapting the level of difficulty to the abilities within your class.

Setting the homework
Provide a similar task to the one given on the sheet, for example, making 11, 12, 14, 18 and 21 using 2s, 3s and 5s only. In each case, stress that there is more than one solution.

Back at school
Encourage the children to appreciate that totals can sometimes be made in more than one way. Use the exercise as an opportunity to consider which totals cannot be made.

p37 MAKE 20

MATHS TO SHARE

Learning outcomes
- **Know by heart all subtraction facts for each number to 20.**
- Extend understanding of the operation of subtraction and use the related vocabulary.
- Check subtraction with addition.

Lesson context
Conduct some work with the class on complements for a total of 20. Talk over the various strategies used. Next, turn attention to the idea of subtraction from 20, perhaps using a large number line to visualise this. Ultimately the goal is to know bonds to 20 using either recall or rapid derivation.

Setting the homework
Practise skills of calculation such as counting on; bridging through 10 where appropriate. You may need to explain the set activity (Pelmanism).

Back at school
Discuss strategies for calculation, linking bonds to 10 with those to 20 (e.g. $12 + 8 = 20$ relates closely to $2 + 8 = 10$).

p38 WHAT'S IN A NAME?

PRACTICE EXERCISE

Learning outcomes
- Classify and describe 2-D shapes (triangles), referring to properties (number and relative length of sides).
- Make and describe shapes and patterns.

Lesson context
Work with the whole class on looking at properties of various 2-D shapes such as square, rectangle, circle and triangle. Take care to develop definitions that are technically precise. Focus attention on the different types of triangle and their names.

Setting the homework
Explain that this exercise is essentially a recap of some of the ideas discussed above. Also, the task is not so difficult once one or two of the cards have been grouped. Alert the children to the fact that some of the words may be unfamiliar to some adults.

Back at school
It should be recognised that the families created are not entirely exclusive. A right-angled triangle, for example, might also be isosceles in some cases.

p39 ALL IN SHAPE (1) MATHS TO SHARE.

Learning outcomes
- Classify and describe 3-D shapes, including the hemisphere and prism, referring to properties such as number of faces, edges and vertices.

Lesson context
Present the class with representations of solid shapes, focusing on both properties and technical names. Work in reverse, offering clues about the shape, for the children to tell you what it is. Work in groups could usefully build on grouping similar shapes, creating framework representations, and/or detailing their properties.

Setting the homework
Provide the children with the homework sheet, making it clear what is meant by the heading 'mathematical name'. You should also make it clear that most shapes tend to fall into just a couple of categories (cuboids and cylinders) and that the children will need to be particularly observant if they are to find more than this.

Back at school
It would be helpful to reinforce the technical language of 'face', 'vertex' and 'edge'. This might also be an opportune moment to clarify that a cube is simply a special cuboid.

p40 ALL IN SHAPE (2) MATHS TO SHARE

Learning outcomes
- Classify and describe 2-D shapes, referring to properties such as number of sides and vertices.

Lesson context
Show the class a range of plane shapes, referring to their properties of side and angle. This should lead to a discussion of regularity (see 'Dear Helper' note). Group work could involve the use of squared and/or isometric paper to create irregular shapes of one specific type, for example, pentagons.

Setting the homework
Alert the children to potentially rich sources for shapes, and the fact that the search involves looking at the faces of solid, often bulky, shapes.

Back at school
Share findings, especially those featuring shapes of many sides. Use this as an opportunity to assess the understanding of the term 'regular'.

p41 LOOK AT THE LABEL MATHS TO SHARE

Learning outcome
- Know the relationship between kilograms and grams.

Lesson context
Provide a range of food packages and demonstrate how to measure their respective masses using suitable measuring equipment such as kitchen scales, spring balance or pan balance. Sort the children into working groups. Each group measures the mass of each item and records this information pictorially or in words, in ascending order, starting with the lightest.

Setting the homework
Explain that this homework requires them to look closely at packaging to find information on the mass contained. It will be helpful to reaffirm the relationship between kilogrammes and grammes, and to alert the children to the fact that some products also provide imperial equivalents.

Back at school
Talk about the different items identified by individuals, looking particularly at any issues arising from the inter-relationship between the metric units and comparative imperial units. *Were any items particularly heavy or light?* Consider what a kilogram of a chosen item would look like (a kilogram of cotton wool would go a long way!).

p42 HOW LONG? MATHS TO SHARE

Learning outcomes
- Begin to use decimal notation for metres and centimetres.
- Measure and compare using standard units (m, cm).
- Suggest suitable units and measuring equipment to estimate or measure length.
- Read scales to the nearest division (labelled or unlabelled).

Lesson context
Provide a range of items for measurement and demonstrate how to measure their respective length and/or width using suitable measuring equipment such as a ruler or tape measure. Sort the children into working groups. Each group measures the length of each item and records this information pictorially or in words, in ascending order, starting with the shortest. One issue you may need to address is what constitutes the item's length? Generally speaking, an item's length is the measurement of its longest side.

Setting the homework
Use this section as an opportunity to reaffirm the conventions for recording length. For example, 1 metre 28 centimetres is 1.28m or 128cm. Talk about the different measuring devices that may be available to the children at home. Be prepared to supply suitable devices such as flexible measuring tapes for some individuals.

Back at school
Look at the different ways individuals have recorded their work, using these examples to reinforce consistent approaches to communicating length. Consider also the different measuring instruments.

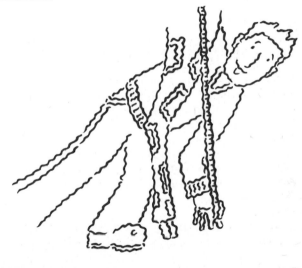

p43 MORE MEASURES MATHS TO SHARE

Learning outcomes
- Begin to use decimal notation for grammes and kilogrammes.
- Measure and compare using standard units (g, kg).
- Suggest suitable units and measuring equipment to estimate or measure mass.
- Read scales to the nearest division (labelled or unlabelled).

Lesson context
This session is as described for page 41, 'Look at the label'.

Setting the homework
Use this time as an opportunity to reaffirm the conventions for recording mass, for example 250g or 0.25kg. Talk about the different measuring devices that may be available to the children at home, but be aware that not all homes will have equipment, such as kitchen scales, necessarily.

Back at school
Look at the different ways individuals have recorded their work, using these examples to reinforce consistent approaches to communicating mass. Consider also the different measuring instruments.

p44 LONGER – SHORTER

Learning outcomes
- Begin to use decimal notation for metres and centimetres.
- Measure and compare using standard units (m, cm).
- Suggest suitable units and measuring equipment to estimate or measure length.
- Read scales to the nearest division (labelled or unlabelled).

Lesson context
This session is based on that described for page 42, 'How long?'.

Setting the homework
Provide flexible measuring tapes where necessary, explaining that this activity will involve measurements up to and beyond 1 metre.

Back at school
Talk about the measurement of flat and curved lengths, and the different items best suited for measurement of these. Confirm that the children are recording correctly in relation to the conventions of length.

p45 HOW MANY?

Learning outcomes
- Count larger collections by grouping them: for example, in tens, then other numbers.
- Give a sensible estimate of up to 100 objects.

Lesson context
Talk with the class about estimation of a large number of objects. Look at large collections of objects or a crowd scene. Discuss strategies for estimation, such as visualising ten and then 'multiplying up'. You might link this with strategies for counting a large pile of coins: *How many pound coins would be needed to create a stack taller than yourself?* Provide groups with large quantities of counting objects, encouraging them to count in ones, twos, by tallying in fives and so on.

Setting the homework
Explain that the homework is a direct extension of the task undertaken in class. Discuss what collections individuals might have for counting. Stress the importance of estimation and the fact that it is acceptable for the estimate to differ from the actual number.

Back at school
Use this as an opportunity to share findings and to revise work on tens: *How much is 20 tens?* or *How many 10 pence coins make a total of £1.30?*

p46 SEQUENCES (1)

Learning outcomes
- Describe and extend number sequences; count on or back in tens or hundreds, starting from any two- or three-digit number.

Lesson context
Work with the whole class on extending simple sequences (e.g. 1, 3, 5, 7... and 3, 13, 23, 33...). Try in particular to sample sequences involving addition of 1, 10 and 100, discussing how this impacts on the sequence of digits. Provide groups with suitable start numbers and 'step sizes'. The use of calculators is optional.

Setting the homework
Explain that the homework samples the idea of counting back in tens and (possibly) in hundreds. Model the activity using your own set of cards.

Back at school
Draw attention to the place value aspects of this activity, clarifying which digits change and which digits remain constant.

p47 PIGS AND DUCKS

Learning outcomes
- Solve mathematical problems or puzzles, recognise simple patterns and relationships, generalise and predict. Suggest extensions by asking 'What if... ?'

Lesson context
Provide groups with target numbers (e.g. 15, 18, 7, 21, 13) and two digits (e.g. 4 and 5). The task is for individuals or pairs to make the target totals by combining the two digits in different proportions (e.g. $15 = 5 + 5 + 5$, $13 = 4 + 4 + 5$). Afterwards, talk about the totals that were/were not possible.

Setting the homework
This task is essentially an extension of the lesson activity. You might show the children one combination without allowing them the opportunity to record this solution.

Back at school
Talk to the children about their strategies for finding the solutions and, most importantly, whether they found them all.

p48 TARGET 80

Learning outcomes
- Solve mathematical problems or puzzles, recognise simple patterns and relationships, generalise and predict. Suggest extensions by asking 'What if... ?'

Lesson context
Working individually or in pairs, the children should investigate totals when three consecutive numbers are added (e.g. $2 + 3 + 4 = 9$). After some time for investigation, draw the class or group together to consider the odd/even nature of answers, and whether the wide generalisation has been spotted (i.e. the answers should all be multiples of 3).

Setting the homework
This homework samples the same objectives and calls upon skills such as trying different combinations. The task is, however, somewhat different and will require some explanation. Use the examples provided to clarify the task and to set the challenge.

Back at school
Look at the range of solutions comparing, for example, whether a number just below 80 is 'better' than one just over 80. Identify collectively the 'best' possible solution: $1 + 23 + 4 + 56 = 84$.

p49 SUPERBUGS!

Learning outcomes
- Understand multiplication as repeated addition. Read and begin to write the related vocabulary.
- Extend understanding that multiplication can be done in any order.

Lesson context
Work with an OHP and 12 counters to create rectangular arrays (e.g. 2 rows of 6). Can the children suggest others? (including 1×12). Emphasise the commutative aspect of multiplication; that is, that 3 rows of 4 is equivalent to 4 rows of 3. Set the children working individually or in pairs on a similar task involving 24 counters.

Setting the homework
This activity extends the lesson idea by looking at several sets of objects (repeated addition) in a different context. You will need to prepare the activity sheet by adding an appropriate number of legs to suit the needs of the class, group and/or individuals.

Back at school
Review the homework and then extend this by asking questions such as: *18 legs, how many bugs?*

p50 SUPERSHAPES!

Learning outcomes
- Understand multiplication as repeated addition. Read and begin to write the related vocabulary.
- Extend understanding that multiplication can be done in any order.

Lesson context
This session builds on the previous activity involving 24 counters, working with the class to consider the factors of 12. These can be represented in the form of a spider, with the number and its associated factors on the body and legs respectively. The children should work on the task as individuals or small groups.

They should see which numbers are factors of 24 and record these in the way suggested by the illustration above. Once finished, the children could use other numbers for investigation.

Setting the homework
This homework further consolidates 'Superbugs!' on page 49, this time using the context of shape. As before, you will need to insert the shape in the first row according to the needs of the individual or group, before copying the homework sheets.

Back at school
Focus on recall of table facts. *How many corners do 6 hexagons have?* Also, investigate whether any of the unit's digits begin to form a repeating sequence. In the case of the hexagons' pattern, for example, the sequence of the units digit should be: 6, 2, 8, 4, 0, 6, 2, 8, 4, 0 – i.e. a pattern of five digits repeated twice.

p51 FAIR SHARES

Learning outcomes
- **Understand division** as sharing. Read and begin to write the related vocabulary.
- Check division with multiplication.

Lesson context
Working with the class in a similar way to the previous two sessions, explore division of 12 by physically sharing 12 counters (possibly on an OHP) between 1, then 2, 3 and so on. For each sort, focus in on those where the outcome results in a fair share with all the counters used (i.e. the factors of 12). Move into group work, providing a suitable start number such as 24. Ask the children to record their 'successful' solutions, possibly as suggested below:

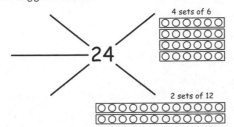

Setting the homework
The homework reinforces the activity undertaken in the lesson. With the whole class, you might demonstrate the task with a chosen number and use this to further clarify the tabular recording. Some children may benefit from some carefully chosen numbers entered on the sheet in advance.

Back at school
Gather a list of those numbers that could be divided by 2, 3 and 4 simultaneously (these should be the multiples of 12). With a confident group you could introduce a Venn diagram with three intersecting sets, entering some numbers in the appropriate regions.

p52 MAKING MONEY

Learning outcome
- Recognise all coins.

Lesson context
As a whole class, discuss different ways of making 5p. In each case consider how many coins are used, whether one or two coin types are represented and whether a different order constitutes a different arrangement. Organise the class into working groups and ask them to find different ways of making 10p or other suitable totals.

Setting the homework
Explain that the homework extends the lesson and is best conducted with a small collection of real coins. Select a target total and model the task so that children are clear what to do.

Back at school
Select a range of examples to try to encompass the full range of coins. You could also use this opportunity to show the amounts in the conventional ways (e.g. £1.20 or 120p).

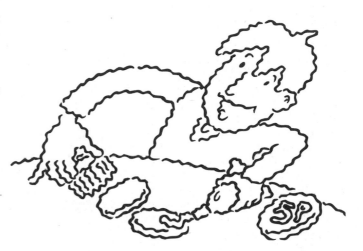

p53 WHAT SHALL I BUY?

Learning outcomes
- Recognise all coins and notes.
- Solve problems involving money. Explain how the problem was solved.

Lesson context
Working with the whole class initially, engage in some simple shopping scenarios requiring a single purchase and giving change. Use this as an opportunity to develop the language of shopping. Move into group work with similar tasks to model in pairs. Alternatively, use worksheets with practice questions involving problem solving with money. Provide real coins.

Setting the homework
You will need to prepare the sheet in advance with appropriate amounts for the three items. Able children should be given quantities that give totals well beyond £1. Emphasise the importance of being thorough and systematic if they are to find all possible answers.

Back at school
Select a range of examples to try to encompass the full range of coins. You should also use this opportunity to show the amounts in the conventional ways (e.g. £1.20 or 120p). In all cases there should be six possible combinations. With products a, b and c: a + b, a + c, b + c, a + a, b + b, c + c.

p54 SHAPE SHARE — PRACTICE EXERCISE

Learning outcomes

- **Recognise unit fractions such as $\frac{1}{2}$ and $\frac{1}{4}$ and use them to find fractions of shapes.**

Lesson context

Talk to the whole class about halving shapes and numbers. Create a large 4 × 4 square and ask for different ways of dividing the shape in half using a single unbroken line. Set working groups on this task, using squared paper for recording. The line can be straight or curved. Furthermore, the line can 'walk a path' which meanders to form two dissimilar shapes, but with the same area of 8 square units.

Setting the homework

The task should be largely self-explanatory. Talk about the importance of counting the number of segments in each shape, to help with the calculation of a half and a quarter. The intention is that segments will not be half-filled, something that can happen when children are presented with some images which stray from the conventional horizontal–vertical conventions.

Back at school

Build on the equivalence aspect, arising from sub-dividing the shapes: $\frac{1}{2} = \frac{2}{4} = \frac{4}{8} = \frac{6}{12}$; $\frac{1}{4} = \frac{2}{8} = \frac{3}{12}$.

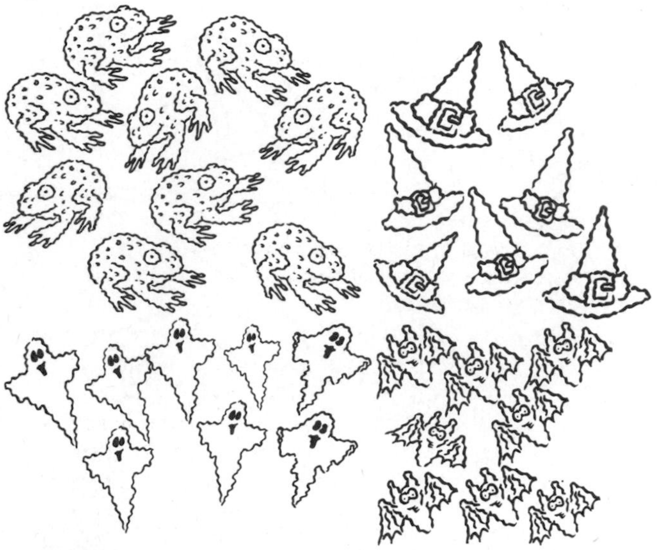

p55 DIVIDING OUT — PRACTICE EXERCISE

Learning outcomes

- Begin to recognise simple equivalent fractions.
- **Recognise unit fractions such as $\frac{1}{2}$ and $\frac{1}{4}$, and use them to find fractions of shapes and numbers.**

Lesson context

With the whole class, review the application and meaning of area. Talk about area in terms of the coverage of a surface. Consider the unit of this measure (square centimetres, square metres). The task is to draw a selection of rectangles (including squares) and to divide each one into two halves, shading one half each time. For each shape, write the area of the whole shape and the area of the shaded half.

Setting the homework

This builds on the previous homework, this time focusing more explicitly on a fraction of a quantity rather than the sub-division of a whole. Emphasise the importance of sharing into equal subsets to generate halves and subsequent quarters as 'half of a half'.

Back at school

Check through the answers and begin to consider the idea of finding $\frac{3}{4}$, by repeated addition of a quarter and by adding $\frac{1}{4}$ to $\frac{1}{2}$. You might also investigate the potential for other fractions such as $\frac{1}{8}$ of 16 or $\frac{1}{3}$ of 24.

p56 PART SHARES — PRACTICE EXERCISE

Learning outcomes

- **Recognise unit fractions such as $\frac{1}{2}$ and $\frac{1}{4}$ and use them to find fractions of shapes and numbers.**
- Begin to recognise simple fractions that are several parts of a whole, e.g. $\frac{3}{4}$.

Lesson context

Ask the class some questions involving practical applications of fractions: *If half a class of 36 is made up of boys, how many girls are there in the class?* Ask similar questions, extending to those involving $\frac{3}{4}$ of a quantity. Undertake group work with some prepared written questions involving fractions:
A bus holds 20 people. If $\frac{1}{4}$ get off, how many are left on? What is $\frac{1}{4}$ of 28? Is it better to have $\frac{1}{2}$ of 16 or $\frac{3}{4}$ of 12? What is half of £1.28?

Setting the homework

This is essentially an extension of the previous homework, 'Dividing out', and the lesson detailed above. Working with the whole class, use the strategies from these sessions to find $\frac{3}{4}$ of a quantity.

Back at school

Check for any significant problems and encourage children to share their strategies. Develop these ideas further by setting similar problems involving thirds, fifths and so on.

p57 JUMPS — PRACTICE EXERCISE

Learning outcome

- Use informal pencil and paper methods to support, record or explain addition.

Lesson context

Group the children according to ability and provide each table with suitable 'start numbers' and 'jump sizes'. These can be written on cards to offer a selection for mixing in different combinations. Children should record their answers as 'jumps' along an empty number line.

Setting the homework

This homework further supports the idea of addition, this time using a number grid. Begin by showing movement on a large grid, perhaps beginning with 'near multiples of 10' such as 9, 19 or 11. Explain how movement down a column changes the tens digit and a row movement changes the units/ones.

Back at school

Discuss whether advanced strategies were used; namely the idea of 18 being two column movements (20) less two row movements left (– 2). The resulting movement is thus 18 units. Use this strategy for larger near multiples of 10 such as 39 and 58.

p58 GIANT JUMPS — PRACTICE EXERCISE

Learning outcome

- Use informal pencil and paper methods to support, record or explain addition.

Lesson context

Provide an empty number line alongside a linear statement. Enter the start number to the left hand edge of the line and then the subsequent 'stopping points' determined by the number of tens and ones to be added.

Setting the homework

The homework uses the same empty number line approach but explores the idea of 'bridging through 10'. Try to incorporate this into the lesson so that the idea is familiar to the children.

Back at school

Review and extend the work, possibly with examples that stretch beyond 100.

Learning outcomes
- Understand and use the vocabulary related to time.
- **Use units of time and know the relationships between them.**

Lesson context
With the whole class standing, challenge the children to estimate one minute simply using their powers of estimation. They should sit down when they think one minute has passed. Next consider the current time to the nearest quarter hour, recording in both digital (not 24 hour) and analogue formats. Use a geared clock (if available) to demonstrate other times, including quarter-past and quarter-to the hour. Engage in various group activities involving the use of timers; for example, reading a page of a book, or doing step-ups. Ask the children to find ways of recording their group activities, possibly using a tabular format.

Setting the homework
The homework further supports the recording of time. Explain that they can either enter the times to the nearest quarter hour or (particularly if support is available to them) to the nearest 5 minutes. You can encourage the children to add the digitally equivalent time also.

Back at school
Look at some of the work collectively, refining any of the conventions that may have been missed. You might also introduce the idea that neither the digital nor analogue methods used indicate whether the timing is am or pm. These abbreviations could be added wherever appropriate.

Learning outcome
- **Solve a given problem by organising and interpreting numerical data in simple lists, tables and graphs.**

Lesson context
In this session a single set ring is introduced. Explain to everyone the idea of negation; that is, that everying inside the ring meets the given criteria and everything outside does not. Provide each group with a related task, appropriate to their current learning needs. These might relate to further tables practice or shape work. You will need to provide numbers or shapes to enable the children to sort onto the sorting boards.

Setting the homework
This covers the same objective as the lesson but involves the collection of data from what should be readily available listings. Television viewing is not necessarily encouraged (or available) in every household. Low frequency of viewing does not negate this task, as the children's entries are essentially a 'wish list'.

Back at school
Gather feedback from the children and begin to formulate categories for different types of programme. This is a theme developed in the homework task 'TV (3)' on page 62.

Learning outcome
- **Solve a given problem by organising and interpreting numerical data in simple lists, tables and graphs.**

Lesson context
The lesson takes the sorting idea a stage further, this time making use of a two-way sort on a Carroll Diagram. Present the class with a giant version of this diagram: Using a range of coloured plane shapes, discuss in

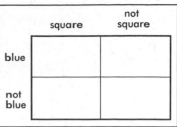

	square	not square
blue		
not blue		

which of the four regions each shape should be placed. Provide similar group tasks, possibly involving the sorting of numbers. For example, you could even use criteria 'even/not even' or '10 or more/fewer than 10'.

Setting the homework
This homework extends the previous homework exercise, requiring additional information to be added. Talk to the group about the idea of time differences, encouraging children to seek help with this at home.

Back at school
Look at the variations in time spent viewing, both between children and across the week. Consider the reasons for such differences.

Learning outcome
- **Solve a given problem by organising and interpreting numerical data in simple lists, tables and graphs.**

Lesson context
Introduce the idea of an individual survey. Draw up a three-column chart. Add suitable entries in the left-hand column such as favourite books/bands/lessons. Provide each child with off-cuts of paper on which to write their favoured choice, avoiding collaboration to give a 'secret ballot'. Collect these in and ask a caller to rapidly call out the choices for another child to tally in the middle column of the chart. Total in the right-hand column, explaining that this constitutes a frequency chart. If time allows, engage groups of children in group surveys on pre-selected themes.

Setting the homework
Refer back to the follow-up discussion detailed in 'TV (1)', regarding the different types of programme. Explain that this exercise requires the help of friends and family members to survey their programme preferences. For simplicity, children must be encouraged to avoid creating other programme categories.

Back at school
Consider whether there seems to be a trend towards any particular programme type. *Might the choice be related to different age groups?* If the children have not already done so, demonstrate how to summate the survey by totalling each column. The category with the lowest total is the overall favourite. *How might a group collate all their results into one presentation to find an overall 'winner'?*

p63 GOING UP

PRACTICE EXERCISE

Learning outcome
• **Order whole numbers to at least 1000,** and position them on a number line.

Lesson context
Initially as a class, 'mix and match' large digit cards (e.g. 3, 6 and 8) to create two- and three-digit numbers (e.g. 63, 836). After recording three or four suitable combinations, re-arrange these in ascending numerical order. Set a similar task with small digit cards, with groups working collaboratively.

Setting the homework
Although the task differs somewhat to that covered in the lesson, the homework activity is essentially one of ordering. Talk through the challenge, clarifying that the ultimate solution is to crack the code.

Answer
Numeracy can be fun!

Back at school
The code-cracking exercise can usefully be applied in other contexts such as computation. Engage in some ordering using negative numbers, larger numbers and/or fractions.

p64 SCALES
MATHS TO SHARE

Learning outcome
• Read scales to the nearest division (labelled or unlabelled).

Lesson context
Present some representations of scales found on a range of measuring equipment found in school and/or on vehicles. Talk about the way in which some dials have all the units marked, whilst others are in steps.

Setting the homework
Consider the health and safety issues of seeking out dials and scales on vehicles, electrical equipment and so on. The children should be encouraged to show some of the details on the scale.

Back at school
Review the range of representations presented. Provide some complex ideas, for example when markings are designed to represent part units, such as the markings between kilograms on bathroom scales.

p65 MAKE IT UP

PRACTICE EXERCISE

Learning outcome
• Extend understanding of the operation of addition and recognise that addition can be done in any order.

Lesson context
Conduct a class discussion on the language of addition. The questions might be presented in a written and/or oral format. The focus of your exposition should be to present the 'addition-related' words featured on the homework sheet. Follow up your discussion with a suitable range of prepared questions such as: *What is the total of 27 and 7? What must be added to 12 to make 21?*

Setting the homework
Read through the task on the homework sheet and talk over the example provided. Check for understanding of the remaining terms. Encourage more able children to create suitably challenging number stories for themselves.

Back at school
Select key words one at a time, and invite volunteers to read out the appropriate sentences. Look at some of the answers featuring several digits and discuss strategies for calculation using informal pencil and paper methods.

p66 NUMBER PAIRS
PRACTICE EXERCISE

Learning outcome
• Extend understanding of the operation of addition and recognise that addition can be done in any order.

Lesson context
Working with the class initially, present a set of five numbers (e.g. 10, 21, 8, 6 and 3) and invite different combinations by selecting any two numbers for addition. These totals could be calculated as you progress through the activity and marked off on a number grid.

Setting the homework
Talk through the homework sheet. The exercise is an extension of the one covered in the lesson. You may want to modify some/all of the numbers to make it easier/harder for individuals/groups.

Back at school
Discuss how many different combinations were found, pointing out that the use of the word 'different' depends on how we interpret the challenge. If we allow an addition pair to be reversed, e.g. 18 + 24 and 24 + 18, there will be 12 combinations. If we do not, there should only be six. What would happen if we also allowed each number to be added to itself?

p67 MAKE IT PAY (2)

PRACTICE EXERCISE

Learning outcomes
• Recognise all coins (to 50p).
• Solve problems involving money, including finding totals and working out which coins to pay.

Lesson context
Working with the whole class initially, recap on the different coin denominations to at least 50p. Practise adding up two or three amounts and giving the appropriate change for £1.00. Set a task where the aim is to make a range of 'three-item' meals and to find the cost for each.

Setting the homework
Provide totals up to the value of 50p and, for each case, identify which coins would be used to make that total. Use could be made of enlarged versions of coins, or OHP versions. Make links with the homework task. You will need to prepare some numeral cards from 11–19 inclusive, or ask children to create these for themselves from scraps of paper or card.

Back at school
Consider which amounts required the most coins. The number of coins required for any total to 50p rarely needs to exceed five.

p68 MAKING MORE MONEY
PRACTICE EXERCISE

Learning outcome
• Recognise all coins.

Lesson context
If possible, provide giant coins or acetate copies of the coins for use on the OHP. Ask some questions involving equivalence such as: *Can you think of different ways of making 20p?* Move on to larger quantities if possible. Organise children into working groups. Ask them to find different ways of making a given amount, recording their answers in a pictorial way.

Setting the homework
Explain that there are sufficient coin illustrations to make all the required totals simultaneously. It is important, however, that the tiles are not glued down until all combinations are found.

Back at school
Check that children have understood the idea of selecting the highest possible denomination at each stage of making any given amount. Follow up with work on giving change from £5.00, this time modelling the 'shopkeeper's method' of counting on from the cost of the item.

p69 ONE STOP PARTY SHOP!
MATHS TO SHARE

Learning outcomes
• Solve problems involving money in 'real life'. Explain how the problem was solved.

Lesson context
Working with the whole class, create two/three problems involving computation in money. For example: *£2.50 pocket money per week, how long to save £30.00?* or *An adult swim is £2.80, a child swim is half that amount. How much for an adult and a child?* Provide group tasks based on questions of this type.

Setting the homework
Although the activity is closed and relatively self-explanatory, it is worth encouraging the children to seek support if it is available at home. The multi-step nature of the task makes it likely that errors will be made. You could simplify the task by reducing the size of the party group, the number of items to purchase and/or the unit costs.

Back at school
Consider the different approaches to calculation such as finding the total cost for one person and then 'multiplying up'. Alternatively, increase the unit costs first and then find the sum of all these sub-totals. You might also discuss the convenient pairings of items to make calculation easier.

p70 WORD PROBLEMS!
PRACTICE EXERCISE

Learning outcomes
• Solve problems involving numbers in 'real life'. Explain how the problem was solved.
• **Choose and use appropriate operations (including multiplication and division) to solve word problems.**

Lesson context
Set some problems orally for the class such as: *I think of a number and subtract 42. If the answer is 16, what number was I thinking of originally?* Try to ask questions encompassing all number operations and, where possible, include some contextualised problems in money and measures. Provide similar problems for small groups to work on as a collaborative exercise.

Setting the homework
Essentially the questions should be consolidatory in nature, based wholly on the ideas developed in the lesson. The idea of showing all working may need some emphasis, as children may be resistant to exposing their informal jottings when they have the opportunity to 'polish' their work.

Back at school
Use each question as an opportunity to share different strategies. Encourage children to show their techniques on the board, in order for others to evaluate their ideas.

p71 SHAPES IN WORDS
PRACTICE EXERCISE

Learning outcome
• Make and describe shapes and patterns.

Lesson context
Working with the whole class, move five cubes on an OHP to create different images. All five cubes must be used each time. At least one face of the cube must be matched with another. Explore one or two shapes, explaining that the projected images are known as 'pentominoes'. Working in groups, ask the children to record all 12 shapes on squared paper.

Setting the homework
Show the children representations of the shapes detailed on the homework sheet. Talk through the properties of each one in turn, or give clues for children to match to the shapes. Provide copies of the homework sheet and explain the task.

Back at school
Use the review of this work as an opportunity to make links between the various shapes, such as: the shapes are all quadrilaterals except the triangle; all of the featured quadrilaterals are types of parallelogram; the square is a 'special case' rectangle.

p72 SYMMETRY
MATHS TO SHARE

Learning outcome
• **Identify** and sketch **lines of symmetry in simple shapes, and recognise shapes with no lines of symmetry.**

Lesson context
Discuss the idea of line symmetry as a class. Using an OHP, lay six connected cubes flat on the glass surface in such a way that the image is symmetrical. Invite individuals to create one or two more, each time using all six cubes. Set the same task for children to work on and record in pairs.

Setting the homework
The lesson has covered aspects of symmetry and asymmetry. Talk about the types of places symmetry might be found, perhaps considering both natural and manufactured items.

Back at school
Share examples found around the home, giving due attention to the mathematical qualities that give the items symmetry.

p73 ABOUT TIME
MATHS TO SHARE

Learning outcomes
• Read and begin to write the vocabulary related to time.
• **Use units of time and know the relationships between them (day, week, month, year).**

Lesson context
Conduct a general class discussion on the language of time. Focus on the units, beginning with the second and expanding this up to one year and beyond. Bring out the irregularities in the system. Working in groups, ask the children to write down the months of the year and the associated number of days in each month. Those children who finish early should calculate the number of days to a significant school or personal date.

Setting the homework
This is essentially a fact-finding exercise that leads to further use of units of time. Explain that they will need to gather information from family members or friends.

Back at school
Make use of an enlarged calendar featuring one month of the year. Look at the features: months do not always start on a Monday; the length of the month; the use of column and rows to find, for example, the dates of Wednesdays in that month; finding differences between dates.

p74 PICTURE THIS — PRACTICE EXERCISE

Learning outcome
* Investigate a general statement about familiar shapes by finding examples that satisfy it.

Lesson context
Revise the properties of various four-sided shapes (quadrilaterals). Introduce the idea of a branching diagram, asking a 'yes/no' question to separate the shapes out into subsets. This branching can be continued with follow-up questions. Work with the whole group on creating a giant floor version or engage the children in working on prepared branching questions.

Setting the homework
This is an open activity to practise recognition, classification and construction of shapes. The children will need to use a ruler. Ensure that children are familiar with colour coding. This is similar to work involving the use of a key.

Back at school
Share the outcome of this exercise. Use an enlarged grid, or OHP of same, to construct other shapes featuring 3, 4, 5, 6 sides. Distinguish between regular and irregular shapes.

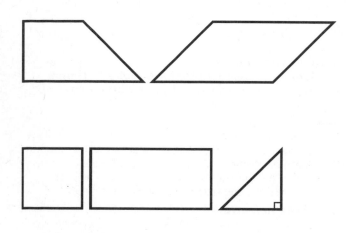

p75 BOXES — GAMES AND PUZZLES

Learning outcomes
* Make and describe shapes and patterns: for example, explore the different shapes that can be made from four cubes.
* Classify and describe 3-D and 2-D shapes.

Lesson context
Create four small card triangles, by cutting across both diagonals of a (square) piece of card. Lay all four shapes together, edge to edge, to create and name the following: square, rectangle, triangle, parallelogram (no right angles). Provide pairs of children with a set of four triangles to create, record (on plain paper) and label the shapes discussed above. Those finishing early should create irregular shapes with five, six or more sides.

Setting the homework
This investigation involves problem solving. Encourage the children to work in a systematic way by, for example, beginning with horizontal and vertical arrangements to identify the two extremes. All the other shapes must be either two or three blocks in height. There are eight solutions in total.

Back at school
Extend the task by working collectively on arranging five cubes using the same rules/constraints. There are 16 solutions possible. Work in a systematic way by, for example, identifying the arrangements reaching one block high (1), then two blocks high (7), three blocks (5), and four blocks (2) and five blocks (1).

p76 ON THE GRID — PRACTICE EXERCISE

Learning outcomes
* Read and begin to write the vocabulary related to position, direction and movement.

Lesson context
Use an 8 × 8 grid, enlarged on an OHP. Provide some pre-prepared coordinates, so that when these are plotted and joined, a simple outline picture is created. Emphasise the reference to horizontal before vertical coordinates. Sort groups to work on a picture of their own, with either coordinates prepared in advance or with children creating a simple picture and then 'converting' this into a set of ordered pairs.

Setting the homework
Emphasise the idea of a pattern forming and stress that it should be immediately apparent if there has been an error in mapping one point.

Back at school
Talk about the enlargements created, possibly using an enlarged version of the complete task. Focus on the area of each square, leading to a recognition that an enlargement of scale factor 2 and 3 leads to a four-fold and nine-fold increase in the respective areas.

p77 GET INTO SHAPE — PRACTICE EXERCISE

Learning outcomes
* Classify and describe 2-D shapes, referring to their properties.

Lesson context
Provide the class with a list of shapes, representing the range of plane (2-D) shapes covered to date. Talk through the properties of some of the shapes, drawing particularly on the precise mathematical vocabulary. Provide groups with paper and set the task of drawing and describing in words a selection of shapes. Some children may benefit from a simpler matching exercise between shape pictures and shape names.

Setting the homework
This homework involves developing children's knowledge of 3-D shapes. Explain that the isometric grid is designed to aid the construction of representations of these shapes.

Back at school
Look at the work in terms of each shape's number of sides, edges and vertices. You may wish to draw the children's attention to the fact that the sum of the faces and vertices is 2 more than the number of edges, in each case.

p78 SEQUENCES (2) — PRACTICE EXERCISE

Learning outcomes
* Count on or back in twos starting from any two-digit number, and recognise odd and even numbers to at least 100.

Lesson context
Ask randomised questions involving knowledge of 5 times table facts. Record the answers in two columns as numbers ending with a 0 or with a 5. Investigate the final digit created by adding on 2 repeatedly from an even number, and then from an odd number. Break the children into groups to look at patterns involving, for example, constant addition of 3.

Setting the homework
Explain that this homework is about completing patterns of constant addition or subtraction. Clarify that the missing number cannot simply be any number that falls between its two neighbours, it must continue the constant rule.

Back at school
Review the solutions and move on to sequences involving fractions or negative numbers.

p79 SEQUENCES (3) PRACTICE EXERCISE

Learning outcome
- Count on in steps of 3, 4 or 5 from any small number to at least 50, then back again.

Lesson context
Create sequences involving the constant addition of 3, 4 and 5, illustrating each on a number line. The follow-up task involves groups creating different-shaped grids containing the numbers 1–36 (e.g. 9 × 4 or 6 × 6 squares) and using these to record starting on a single digit number, constantly adding on 3 or 4 or 5, and observing whether this produces a regular pattern. Look at the units digits of each sequence to see when these repeat.

Setting the homework
This is essentially an extension of the previous 'Sequences (2)' homework. As before, it is important that the missing numbers continue the established constant addition/subtraction rule.

Back at school
Extend the sequences beyond what was set as homework, in order to see the cyclical nature of the units digit in each pattern. You should note, for example, that the cycle of repetition is more frequent for addition of 4 than for addition of 3. Discuss the colour patterns created and ask the children to explain these differences in outcome.

p80 STATEMENTS PRACTICE EXERCISE

Learning outcome
- **Recognise that division is the inverse of multiplication.**

Lesson context
Present a large multiplication grid to explore the links between multiplication and division. Show how, for example, 8 × 4 = 32 and that 32 ÷ 4 = 8. Provide groups of children with cards containing sets of numbers and symbols (e.g. 28, 7, 4, =, × and ÷). The groups should create four related statements, two for multiplication and two for division.

Setting the homework
This homework extends the classroom task by offering more scope for creating correct statements. The task can be set as a challenge to see how many different statements can be found.

Back at school
Focus the discussion on the number of solutions found and whether they have used reversals such as 8 × 4 and 4 × 8.

p81 TAKE BACK PRACTICE EXERCISE

Learning outcome
- **Understand division** as grouping (repeated subtraction).

Lesson context
Present an empty number line, marked 0 at one end and 30 at the other. Demonstrate counting back to 0 in steps of 3. Provide the children with plain paper to investigate other step sizes. Ask: *Which steps, starting from 30, finish on 0?*

Setting the homework
This activity extends the idea of repeated subtraction from a range of different numbers and step sizes. Explain that all the sequences should end in 0. When the diagram is complete, make the link with the number sentences given below.

Back at school
Consider other number facts, offering the multiplication facts and inviting the associated division facts.

p82 DIFFERENCES GAMES AND PUZZLES

Learning outcome
- Extend understanding of the operation of subtraction.

Lesson context
Show the children the shrinking squares investigation, generated at each stage by finding the differences between numbers at adjacent corners.

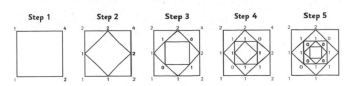

Step 1   Step 2   Step 3   Step 4   Step 5

The steps continue until all the differences are 0. Provide blank paper and some suitable start numbers; the greater the ability, the wider the range and scale of numbers. The task could be organised with groups of three or four children taking turns to generate sequences on large sheets of paper.

Setting the homework
Give out the homework sheets, explaining that the corner numbers are pre-set. Some children may benefit from modified sheets to make the task easier or harder.

Back at school
Discuss some of the patterns arising: *Why do some patterns seem to take more steps than others to complete?*

p83 MONEY PROBLEMS! PRACTICE EXERCISE

Learning outcomes
- Recognise all coins and notes.
- Solve word problems involving money, including finding totals and giving change, and working out which coins to pay. Explain how the problem was solved.

Lesson context
Revisit the different coin denominations, practising counting on from 0 in 2s, 5s, 10s, 20s and 50s. Ask a range of oral questions such as: *How many 20p coins in £3.00?* The children work in groups to represent five of each coin denomination and, in each case, write the total value of the set. Discuss change from £5.00, £10.00 and £20.00.

Setting the homework
The homework gives further practice of calculating totals and giving change. You may need to modify the level of difficulty for some individuals.

Back at school
Focus on strategies as well as answers. When looking at each question, you might brainstorm several different ways of arriving at the correct solution.

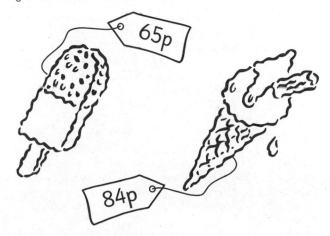

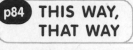

p84 THIS WAY, THAT WAY
PRACTICE EXERCISE

Learning outcomes
• Extend understanding that more than two numbers can be added.
• Use knowledge that addition can be done in any order to do mental calculations more efficiently.

Lesson context
Draw a large square on the board and write a target number inside (e.g. 100). Ask children to volunteer numbers for each corner, the total of which should match the target number. Give the children plain paper or sheets with empty squares drawn on, offer a suitable target number for each broad band of ability and support individuals/groups as necessary.

Setting the homework
Continue the lesson by emphasising how the corner numbers could be added in any order, although on many occasions the numbers lent themselves to addition in a particular order. For example, show how two numbers sometimes conveniently combine to give a multiple of 10.

Back at school
Focus on the language and strategies of calculation. Numbers such as 13 and 17 combine to give a multiple of 10 because 3 and 7 are complements of 10. Other pairs, such as 18 and 12, work because the two numbers compensate one another to give a multiple of 10.

p85 NUMBER MACHINES
PRACTICE EXERCISE

Learning outcome
• Partition into tens and units, then recombine.

Lesson context
Work on complements, giving a multiple of 5 and asking children in the class to 'make up' the number to 100 with the appropriate complement. Use a 100 square, emphasising the idea of breaking the number into tens and units, particularly with those multiples ending in 5 where the tens and units are matched together giving 90 and 10. Practise counting on the 100 grid from any number. Provide some questions such as $63 + \square = 100$, checking as follows:

$$63 + 37 = 60 + 3 + 30 + 7 = 90 + 10 = 100$$

Setting the homework
The homework task further consolidates work on partitioning and recombining.

Back at school
Work on complements to 500 and/or 1000. You might also consolidate this work as movement along a number line, counting on from the first number to the end point.

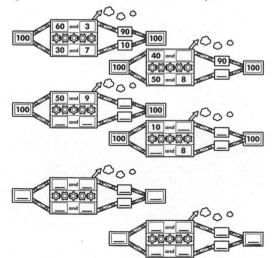

p86 COIN COMBOS
PRACTICE EXERCISE

Learning outcome
• Solve word problems involving money, including finding totals and giving change, and working out which coins to pay. Explain how the problem was solved.

Lesson context
In role of 'shopkeeper', talk about the running of a shop and the skills associated with handling money. Engage children in simple role-play, with some 'set piece' priced items. Next, present three differently-priced bars of chocolate. The 'customer' can make two or more selections, including multiple purchases of one type. Ask the class to help the 'shopkeeper' in calculating the total. Talk through giving change. Ask the children to model the activity in pairs.

Setting the homework
Revise the idea of using the largest-value coin possible at each stage of making up to a given total. Work together on the first example, to ensure everyone has the right idea.

Back at school
Work with larger totals including the use of the £5.00 note. Consider what the 'largest minimum' number of notes and coins might be for any given amount within £10.00 (£9.99 requires a £5.00 note and eight coins).

p87 FRACTION MACHINE
PRACTICE EXERCISE

Learning outcomes
• **Recognise unit fractions such as ½ and ¼, and use them to find fractions of shapes or numbers.**
• Compare familiar fractions.

Lesson context
Talk about repeated doubling starting from 1. Take one of those large numbers and do the reverse (halving back to 1). Provide individuals with 8 x 8 square grids. The children should count all the small squares (64), colour half of them and calculate and differently colour half of the remainder. This continues until only one square is left uncoloured. Finally the number of squares in each colour is noted. Discuss what fraction each colour represents.

Setting the homework
Explain how, in this exercise, tiles are mixed to create number stories involving fractions.

Back at school
Look at the statements created and the way that one solution leads on to another (e.g.: 'quarter of 12 is 3' leads nicely on to 'three-quarters of 12 is 9').

p88 SHAPE SPLIT
PRACTICE EXERCISE

Learning outcomes
• **Recognise unit fractions such as ½ and ¼, and use them to find fractions of shapes.**
• Begin to recognise simple equivalent fractions.

Lesson context
Provide representations of squares divided both across the diagonal(s) and through mid-points of sides, into halves and quarters. Engage groups in cutting up these shapes and using the pieces to recreate unit squares in different ways.

Setting the homework
This homework includes the use of eighths. Although this is unlikely to be new content, it is important to affirm the notion of an eighth as 'half of a quarter'.

Back at school
Extend the idea of equivalence by developing a sequence of fractions with a value of ½ (¾, ⅜, ⅝, ⁵⁄₁₀ and so on). Extend with fractions equivalent to ¼ and ⅛.

p89 ALL IN A DAY INVESTIGATION

Learning outcome
- **Solve a given problem by organising and interpreting numerical data in simple lists, tables and graphs.**

Lesson context
Enter data in a simple graphing program on the computer. This can be generated rapidly by asking for a show of hands based around the choice of favourite food from four options. Look at the different types of graph available, discussing the merits of each. A pie chart works well with the homework below.

Setting the homework
This homework links well with the idea of handling data and the previous two activities involving fractions. Explain that although it is possible to split the sections for times other than 'on the hour', this could make the task unduly complicated.

Back at school
Compare the answers to the questions to try to establish the typical range of responses. Looking at the time spent asleep, use the children's growing understanding of fractional parts to see that this typically equates to more than $\frac{1}{3}$ of a child's day.

p90 BIRTHDAY CHART INVESTIGATION

Learning outcome
- **Solve a given problem by organising and interpreting numerical data in simple lists, tables and graphs.**

Lesson context
Take the class into a large space, selecting a favourite coloured quoit from a choice of three or four options. Use this to create a giant block graph and then, with the class evenly-spaced around an imaginary circumference, a pie chart. If the quoits of the same colour are formed in arcs, the outcome forms the basics of an informal pie chart. Children then work in groups of 8–10 to gather data and record on pre-sectioned circles.

Setting the homework
Gather data on birthday months for everyone in the class, or provide the data from the children's registration details. Talk about the features of the empty bar chart provided.

Back at school
Look at a different bar chart, drawing from numeracy documents or other suitable texts. If possible, look at examples where the y-axis features entries that fall between whole units and/or where the y-axis is labelled in steps other than 1.

p91 PETS CORNER INVESTIGATION

Learning outcome
- **Solve a given problem by organising and interpreting numerical data in simple lists, tables and graphs.**

Lesson context
Discuss questionnaires, in particular the 'yes/no' variety. Select a suitable theme, asking groups to create a suitable set of five yes/no questions. This task will need a lot of management, particularly when the time comes to gather the data. You may elect to spread this task across the week.

Setting the homework
In the same way as for the previous homework, quickly gather the information on pets. The children can then take the key details home to graph the data. Talk through some appropriate mathematical statements that arise from the data, for example: 'There are five more cats than dogs'.

Back at school
Compare this type of graph with the previous example of a bar chart. Consider the main differences and the type of data which is most appropriate to these forms of graph.

p92 HOUSE! GAMES AND PUZZLES

Learning outcomes
- Understand and use the vocabulary of estimation and approximation.
- Round any two-digit number to the nearest 10 and any three-digit number to the nearest 100.

Lesson context
Discuss rounding up and down to the nearest 10. Extend to rounding to 50 and 100. Discuss the use of this in real life, for example, in giving the approximate number of people in a hall. Group the children in pairs with two dice, seven counters and two game boards featuring the multiples of 10 from 10–70. Tell them to take turns and use the dice outcome to create one of two possible numbers. For example, rolls of 3 and 1 can be 31 or 13 to round up or down to nearest 10 and match to an appropriate and vacant multiple of 10 on the game board. First player to cover all their numbers wins.

Setting the homework
The homework activity is similar to that used in the lesson, but is somewhat reversed: the player who gains a multiple of 10 as their outcome has to match this to a number on the game card that has been rounded up/down. You may need to show the children by example.

Back at school
Provide some much larger numbers, to 1000 and beyond. Use these for rounding each to the nearest 10, 50 or 100.

p93 ROUGH TALK PRACTICE EXERCISE

Learning outcomes
- Read and begin to write the vocabulary of estimation and approximation.
- Round any two-digit number to the nearest 10 and any three-digit number to the nearest 100.

Lesson context
Discuss the usefulness of rounding, for example three items at 39p is approximately 3 × 40p. Discuss also how much would need to be added or taken away to 'compensate' for this approximation. Provide groups with number calculations, particularly those with 'near multiples of 10' (e.g. 19 + 42, 6 × 19). Ask the children to approximate using the strategy shown, and to use this to find the 'real' answer.

Setting the homework
This Practice exercise develops the skills in a slightly different context. The children should now be familiar with this 'pick and mix' approach to questions.

Back at school
Review the outcome, paying particular attention to the language of different strategies: rounding, subtract, compensate, approximately, over, under.

p94 STAGING POSTS (1) — PRACTICE EXERCISE

Learning outcomes
- Extend understanding of the operation of addition and recognise that addition can be done in any order.
- Use patterns of similar calculations.

Lesson context
Use a large demonstration number line to travel from any chosen number to 50. Discuss different strategies:
- Start at 26, add 10 (36), add 10 (46), add 4 (50)
- Start at 26, add 4 (30), add 10 (40), add 10 (50)
- Start at 26, add 4 (30), add 20 (50)

In each case emphasise that the overall step is always the same, whatever method is used. Organise children by ability, to practise counting on from numbers of their own choice to an appropriate end-point number such as 50, 100 or 200.

Setting the homework
This homework follows the lesson directly, but limits the children to bridging through 10. If the numbers are too small/large, you may want to modify the sheet for certain individuals/groups.

Back at school
Practise as a class, bridging through 10, 50 and 100, up to 1000. Continue to use the empty number line approach, as this gives good preparation for more formal approaches.

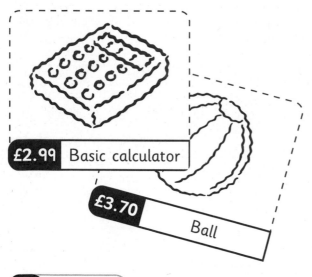

£2.99 Basic calculator

£3.70 Ball

p95 BUDGET — MATHS TO SHARE

Learning outcomes
- Use informal pencil and paper methods to support, record or explain HTU ± HTU.
- **Understand and use £.p notation** (for example, know that £3.06 is £3.00 and 6p).

Lesson context
Use shopping catalogues and advertising leaflets to identify the key elements of their presentation and layout. Provide the children with an imaginary spending limit, which is not to be exceeded. If possible, use leaflets suitable for the children to cut up in order to stick appropriate pictures of their chosen items alongside the prices. The use of calculators is at your discretion. The focus of the exercise is on correct notation of £.p.

Setting the homework
The homework represents a similar exercise to the lesson. The items have been chosen to reflect a range of interests, so everyone should find some appealing purchases. As before, stress that the focus of the exercise is on correct notation of £.p.

Back at school
Look at the amount of change selected individuals had. Discuss strategies for calculating change from £10.00 using counting on. Ensure that the notation meets with general conventions for recording money in figures.

p96 ALL THINGS EQUAL — PRACTICE EXERCISE

Learning outcome
- Extend understanding that subtraction is the inverse of addition.

Lesson context
Discuss the question □ – □ = 10, inviting answers (e.g. 21 – 11, 14 – 4). Extend to double-sided statements (e.g. 10 + □ = 30 – □). Emphasise the = sign as one of equality, rather than simply an instruction to action. Provide different ability groups with suitable open-ended questions of the type shown above. Consider small groups working with large sheets of paper as a collaborative task.

Setting the homework
This homework sheet is designed to consolidate the work in the lesson. Outline the three blocks of tasks, particularly the final one which is open-ended in nature.

Back at school
Look at the range of answers, including the potential for fractions. You could also suggest that the range be extended into negative numbers!

p97 HOW'S THAT? (1) — PRACTICE EXERCISE

Learning outcomes
- Extend understanding of the operations of addition and subtraction; read and begin to write the related vocabulary.

Lesson context
Ask some rapid recall questions involving subtraction. Make sure some of the questions feature missing numbers at places other than the end, for example □ – 12 = 19. Provide groups with a selection of questions, a small space for the answer, and a large space to explain, in writing, how they did it. See the format of the homework for layout.

Setting the homework
This task mirrors that of the lesson, this time using the context of money.

Back at school
Share some of the strategies used including counting on/back, rounding up/down and compensating, and treating the units as '5 and a bit more'.

p98 HOW MUCH? — PRACTICE EXERCISE

Learning outcomes
- Use informal pencil and paper methods to support, record or explain HTU + TU, HTU + HTU.
- Begin to use column addition for HTU + TU where the calculation cannot easily be done mentally.

Lesson context
Talk through the two methods of column addition highlighted on the homework sheet (and in the NNS *Framework*). Ensure that the children are comfortable with both methods before progressing. Organise the children into working groups, providing each with numeral cards: 2, 3, 5, 6 and 7. Using any four cards each time, the children should make different two-digit addition problems and set these out on paper to calculate as they have been shown.

Setting the homework
Encourage the children to find lots of different combinations as this will ensure a range of problems are encountered. Work on HTU is a feature of the follow-up.

Back at school
Talk to the children about the total number of combinations possible. There should be six different combinations, or 12 if 'reversals' are allowed. Practise finding totals involving three-digit numbers if the children are ready to accommodate this.

 p99 STAGING POSTS (2)   PRACTICE EXERCISE

Learning outcome
• Use informal pencil and paper methods to support, record or explain HTU – TU, HTU – HTU.

Lesson context
Following on from the session detailed in the 'Staging posts (1)' homework, practise moving along a number line to find the difference between two numbers. On this occasion, the end-points should be numbers other than multiples of 10. Try to reach a stage where the number of steps can be reduced to three with the staging posts being the next multiple of 10 after and before the first and last numbers respectively. For example, in 64 – 38, add 2 to 38 (40), 20 to 40 (60) and 4 to 60 (64). Provide some similar problems for the class to work with, on their own or in pairs.

Setting the homework
This task is not unlike that set in the 'Staging posts (1)' homework. Tell the children that you want them to move from the first number to the target in fewer moves; no more than two.

Back at school
Provide further practise of finding differences involving three-digit numbers.

p100 GIVING CHANGE   MATHS TO SHARE

Learning outcome
• Recognise all coins and notes.

Lesson context
Revise the 'shopkeeper's method' for giving change. The use of giant coins can be particularly effective on these occasions. Engage the children in paired role-play involving making a total with any two coins, finding and giving the total, and receiving change from £1.00, £5.00 and £10.00, as appropriate.

Setting the homework
This homework samples the aspect of giving change, identifying that the skill of using the 'shopkeeper's method' is fully understood. Encourage the children to work with someone else, and to use real coins if at all possible.

Back at school
Extend work to giving change from £5.00 and £10.00.

p101 PROBLEMS, PROBLEMS!   GAMES AND PUZZLES

Learning outcomes
• Solve word problems involving money. Explain how the problem was solved.
• **Understand and use £.p notation** (for example, know that £3.06 is £3.00 and 6p).

Lesson context
Introduce the class to some closed problems involving money: *A pencil costs 22p and a pen 34p. How much for both? How much more is the pen than the pencil?* Set groups working on problems involving more than one step of calculation: *Two rulers cost 86p. A pen costs 10p more than a ruler. How much do pens and rulers cost?*

Setting the homework
The homework continues the problem-solving theme. Some more/less able groups/individuals may benefit from modifications to the amounts presented on the sheet.

Back at school
Share strategies for calculation, taking each question in turn.

p102 SOLID SHAPES   PRACTICE EXERCISE

Learning outcome
• Classify and describe 3-D shapes (including prisms), referring to their properties.

Lesson context
Use demonstration solid shapes to discuss their properties. (A suitable range of shapes is listed in the Year 3 section of the NNS *Mathematical Vocabulary Book* from the DFEE). Provide the children with a table containing a list of the names of a selection of shapes, or illustrations of them, and three empty columns to enter the number of edges, vertices and faces.

Setting the homework
This task follows neatly on from the lesson, but gathering the same information from objects around the home. To that end, a useful preliminary discussion might consider possible choices of object.

Back at school
Draw the link, made in an earlier follow-up (see 'Getting into shape', page 77), between edges, faces and vertices (the sum of the faces and the vertices exceeds the number of edges by 2). Do the children's findings bear this out?

p103 2-D SHAPES   PRACTICE EXERCISE

Learning outcome
• Classify and describe 2-D shapes, including quadrilaterals, referring to their properties.

Lesson context
Conduct a class revision on the names of 2-D shapes. (A suitable range of shapes is listed in the Year 3 section of the NNS *Mathematical Vocabulary Book* from the DFEE). Introduce the trapezium and parallelogram, if the children are not already familiar with these shape names. Provide each child with a two-piece tangram created from a square cut from one of its corners to the mid-point of the opposite side. Ask the children to make shapes by placing these two cards edge to edge, and to record pictorially, labelling each with the shape's name. Encourage the creation of irregular shapes. Remind the children that any six-sided shape, for example, is a 'hexagon' – regular or irregular.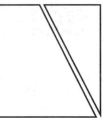

Setting the homework
This is essentially a shape-matching exercise.

Back at school
Review the work. With the children looking at their sheets, the teacher or selected child gives clues about a shape. The class have to put their finger on the shape they think is being described.

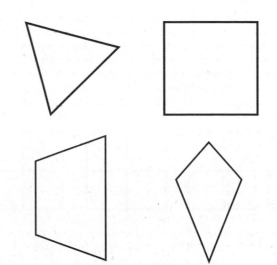

p104 IN BALANCE
PRACTICE EXERCISE

Learning outcomes
- Solve mathematical problems or puzzles, recognise simple patterns and relationships, generalise and predict. Suggest extensions by asking 'What if… ?'
- **Identify** and sketch **lines of symmetry in simple shapes and recognise shapes with no lines of symmetry.**
- Sketch the reflection of a simple shape in a mirror line along one edge.

Lesson context
Review earlier work on line symmetry. As a class complete some 'half pictures' with the mirror line clearly marked. Use an OHP to project cubes arranged as follows:

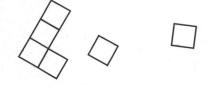

Move the pieces around to connect them in ways that give a symmetrical outline. Provide the children with the cubes to undertake this task for themselves, recording successful arrangements.

Setting the homework
Talk through the example on the homework sheet and work together on drawing another shape that uses the triangles. Mark the line(s) of symmetry.

Back at school
Look at some of the outcome, looking for any answers displaying rotation symmetry but no line symmetry.

p105 SAME AND DIFFERENT
GAMES AND PUZZLES

Learning outcome
- Solve mathematical problems or puzzles, recognise simple patterns and relationships, generalise and predict. Suggest extensions by asking 'What if… ?'

Lesson context
Recap on properties of squares. Outline an investigation involving 9-pin boards and elastic bands: *In how many different places can you create a square? (Six positions.)* Move on to 16-pin boards.

Setting the homework
Use the example on the homework sheet to clarify the task. Emphasise that the edges must be matched and aligned fully, otherwise there would be an infinite number of possible answers.

Back at school
Talk through the notion of 'different'. Identifying some shapes simply as rotations of others will reduce the possible number of different combinations. Care needs to be taken, however, to consider whether a flip movement (reflection) constitutes a different arrangement. Not allowing rotations reduces the number of possible combinations to 7.

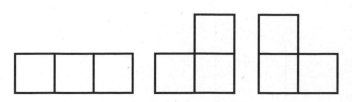

p106 ALL RIGHT NOW
PRACTICE EXERCISE

Learning outcomes
- **Identify right angles** in 2-D shapes and the environment.
- Compare angles with a right angle.

Lesson context
Provide the children with an angle measurer or the corner of a sheet of card with which to test the corners of shapes for right angles. Provide a suitable range of plane shapes and a table headed: 'no right angles', 'all right angles' and 'some right angles'. The children's task is to measure the angles for each shape, and locate the shapes in the appropriate position on the table.

Setting the homework
Explain that the homework provides an angle measurer with which the children can test for right angles at home. Encourage the children to think about suitable surfaces for measuring angles.

Back at school
Discuss whether it was easy to find angles other than right angles (many things are designed with right angles). Introduce the language of 'acute' and 'obtuse' angles to signify angles below 180 degrees other than 90 degrees.

p107 SEEING DOUBLE
PRACTICE EXERCISE

Learning outcome
- Measure and compare using standard units (m, cm).

Lesson context
Talk about scaling up and down, using pictures of microscopic images and maps. Set the children working in small groups of no more than four to create a half-size representation, on paper, of one of the group members. Each measurement should be in proportion. Provide a range of measuring equipment for the children to select from.

Setting the homework
You may need to briefly practise the measurement of small lengths involving millimetres. An OHT provides an excellent model for demonstrating enlargement, simply by moving it away from the screen. Suggest one or two useful items such as a postage stamp.

Back at school
Share the range of items sampled. Consider modelling a ten-fold enlargement using an object that is relatively easy to scale up. With an item such as a stamp, ask: *Does twice as big mean twice the area?* (No, it is actually a four-fold increase.)

p108 TREASURE ISLAND
PRACTICE EXERCISE

Learning outcome
- Read and begin to write the vocabulary related to position, direction and movement.

Lesson context
Gather the children around a giant 2 x 2 grid. Investigate ways of travelling along the lines from the bottom left to the top right. Limit movement to up and right only. *Can all six routes be found?* Provide small representations of the grid, six per child and ask them to highlight the different pathways possible.

Setting the homework
This activity continues the idea of a grid pattern, but picks up on the locational theme developed in Term 2 with coordinates. This is a relatively open and fun activity.

Back at school
Look closely together at a prepared acetate on an OHP of a completed island. Discuss items relative to others using language involving the compass directions, for example: *The caves are north of the swamps.*

p109 IN A SPIN — PRACTICE EXERCISE

Learning outcome
- Read and begin to write the vocabulary related to position, direction and movement.

Lesson context
Use a transparent pegboard and an elastic band on an OHP to create the outline of a simple plane shape. Record this on the board. Rotate the pegboard through a right angle clockwise, recording this to begin a sequence. Continue this and discuss the rotational pattern. Provide groups with dotty paper and pegboards to investigate and create similar patterns.

Setting the homework
This provides a different context for rotational patterns. If possible, provide a representation of the pentagonal example with a split pin through the centre of rotation. Invite individual children to rotate it a quarter turn each and to draw the resulting orientation on the board

Back at school
Talk about turns other than a quarter. Consider also anti-clockwise rotation. Show how a circle does not change under rotation – a unique case.

p110 FULL TO CAPACITY — PRACTICE EXERCISE

Learning outcomes
- Suggest suitable units and measuring equipment to estimate or measure length, mass or capacity.
- Measure and compare using standard units (m, cm, kg, g, l, ml).
- Read scales to the nearest division (labelled or unlabelled).

Lesson context
Ask questions orally in relation to length, mass and capacity. (The NNS *Framework for Teaching Maths* provides some useful examples.) Provide for a circus of activities, possibly spread over several lessons, involving the use of relevant measuring equipment and suitable items to measure. Encourage the children to keep 'field notes' of their work.

Setting the homework
Talk about the different units of capacity (ml, cl and l).

Back at school
Make some introductory links between the capacity of a liquid and its associated mass. The children may be interested to know that under the metric system, 1l of water has a mass of 1kg.

p111 NUMBER SORT — PRACTICE EXERCISE

Learning outcome
- Recognise two-digit and three-digit multiples of 2, 5 or 10.

Lesson context
Return to the idea of developing multiples under repeated addition. Begin with counting in 3s, 4s or 5s and move on to counting in 15s, 20s and 30s. Use one of the earlier patterns, to draw the children's attention to the cyclical nature of the pattern in the units digits. For example, the 4s sequence repeats after every five numbers. Set the children working on less familiar sequences, possibly those relating to 7s, 8s and 9s. The idea is to identify the nature of repetition in these multiples.

Setting the homework
The lesson should have highlighted the fact that some numbers occur in several multiplication patterns; that is numbers with many factors. This homework requires the children to sort numbers into the various regions on a Venn diagram.

Back at school
Focus attention on the numbers in the Venn diagram intersection. These numbers can only be present if they are multiples of 10 and are even. Try with other multiples using the same approach and a different set of numbers for sorting.

p112 HOW'S THAT? (2) — PRACTICE EXERCISE

Learning outcome
- **Explain methods and reasoning** orally and, where appropriate, in writing.

Lesson context
Ask some rapid recall questions involving all four operations. Provide groups with a selection of questions, a small space for the answer and a large space to explain, in writing, how they did it. See the format of the homework for layout.

Setting the homework
This homework is similar to 'How's that? (1)' on page 97. Make clear the importance of providing evidence of strategies.

Back at school
Share strategies for calculation, particularly those involving multiplication and division.

p113 ODDS AND EVENS — GAMES AND PUZZLES

Learning outcome
- Investigate a general statement about familiar numbers by finding examples that satisfy it.

Lesson context
Teach the class how to find the digital root of a number. For example, 27 has a digital root of 9 when both digits are added. Some roots take an additional step (e.g. 48… 12 (4 + 8) …3 (1 + 2)). Select a familiar table pattern and find the digital roots of the sequence. Discuss repetition in the resultant roots.

Setting the homework
This investigation should enable the construction of one or more general statements about odd/even numbers. It may be necessary to attempt to explain further what is meant by the fact that a general statement is one that works for all cases of the same type. For example, 'The 2 times table only features even numbers'.

Back at school
Gather the statements together. Try some examples to 'test' them, writing them in general terms: odd + odd = even, odd + even = odd, and even + even = even

p114 STARTING GRID — GAMES AND PUZZLES

Learning outcomes
- Solve mathematical problems or puzzles, recognise simple patterns or relationships, generalise and predict.
- **Explain methods and reasoning** orally and, where appropriate, in writing.

Lesson context
Working initially as a whole class, attempt to arrange large 1–9 numeral tiles in a 3 x 3 arrangement so that any horizontal, vertical and diagonal line of three numbers gives the same total (15). This can be achieved through trial and improvement, or application of logic, e.g. a 'middle value' number in the middle of the arrangement. Set the same task to the groups or modify to involve any consecutive sequence of numbers (e.g. counting numbers 2–10 or even numbers 2–18).

Setting the homework
This task continues the numbers-in-a-grid approach. As before, answers can be found through perseverance, judgement or a bit of both.

Back at school
Identify some of the internal logic that might have been applied. Often two 'extreme' numbers balance themselves out. On some occasions an even number must be matched with another. These strategies serve to limit the vast number of random possibilities.

p115 WHAT'S THE STORY? — PRACTICE EXERCISE

Learning outcome
- **Choose and use appropriate operations** and ways of calculating **to solve problems.**

Lesson context
Provide some questions on the board and solve them in turn (e.g. 23 + 19 = 42). Work through them again, this time turning each into a contextualised word problem. Organise the children into differentiated groups and provide prepared questions for computation and contextualisation. Offer a layout such as the one suggested on the homework sheet.

Setting the homework
The homework is essentially an extension of the work covered in the lesson. Encourage the children to draw on a broad range of contexts such as money and measures.

Back at school
Select examples which draw on a diversity of mathematical concept areas.

p116 WHAT'S IT WORTH? — GAMES AND PUZZLES

Learning outcomes
- Solve word problems involving money, including finding totals. Explain how the problem was solved.
- **Understand and use £.p notation.**
- Recognise all coins.

Lesson context
Explain to the children that you are holding two coins of undeclared value: *How much might I be holding?* Take a few ideas, recording each amount in pence and £.p notation. Provide the children with paper and coins to investigate the different possibilities.

Setting the homework
Explain that on this occasion the task has many different solutions and it would be unrealistic, therefore, to expect all combinations to be detailed. The important factor in this task is that the amounts are written correctly in the two columns.

Back at school
Look at the least/most amounts possible to establish the range of answers. Next consider amounts below £1.00. Move on to larger amounts, paying particular attention to the inclusion of 0 in amounts such as £3.06.

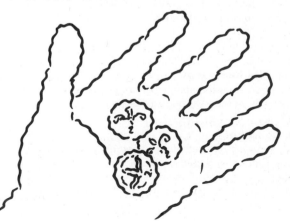

p117 MATHS LINKS — PRACTICE EXERCISE

Learning outcome
- Say or write a division statement corresponding to a given multiplication statement.

Lesson context
Provide some contextual questions involving multiplication/division for example: *How many tandems would you need for eight people?* Stress the inverse nature of multiplication and division. Use a multiplication grid to help establish these links. Set groups a target number (e.g. 4) and ask them to create division facts with that answer (e.g. 24 ÷ 6, 36 ÷ 9).

Setting the homework
Work through the first example, discussing the statements and putting them in real contexts.

Back at school
Look particularly at the last box on the sheet to consider the children's own ideas.

p118 REMAINDERS (1) — PRACTICE EXERCISE

Learning outcomes
- **Understand division** as grouping (repeated subtraction) or sharing.
- Begin to find remainders after simple division.

Lesson context
Revise the idea of a remainder as 'what is left over after sharing'. Use a large space and scatter three PE hoops randomly. Call out numbers between 2 and 6 to denote the size of group of children to be made in each hoop. After each attempt, discuss the underlying mathematics, focusing particularly on any remainders. Allocate the children into groups of five or less at tables. Tell the children to take a handful of counters, share them equally and record the number each person received and note down the remainder, if applicable.

Setting the homework
Encourage the children to use real objects around the home (if possible).

Back at school
Consider which numbers share equally (without remainder) into different numbers of groups (i.e. numbers with several factors).

p119 REMAINDERS (2) — PRACTICE EXERCISE

Learning outcomes
- **Understand division** as grouping (repeated subtraction) or sharing. Read and begin to write the related vocabulary.
- **Recognise that division is the inverse of multiplication.**

Lesson context
Working with the class, write a number such as 2 in the centre of the board and invite divisions that are equivalent. These can be linked directly to the idea of fractions (e.g. $\frac{1}{6}$, $\frac{4}{2}$). Set a suitable target number for each ability group and support as required.

Setting the homework
This activity is essentially the same type of task as the previous homework, but in reverse. This may test the children's understanding more fully. Consider the worked example together.

Back at school
Work through the links between the multiplication tables and the remainders. For example, 13 ÷ 4 can be linked with the idea of 3 x 4 + 1. Note that in the latter statement, brackets are not required as the order of operations is unambiguously favoured to multiplication.

p120 FRACTION PIECES (1)
PRACTICE EXERCISE

Learning outcomes
- **Recognise unit fractions such as ½ and ¼ and use them to find fractions of shapes.**
- Begin to recognise simple fractions that are several parts of a whole, such as ¾.
- Begin to recognise simple equivalent fractions.
- Compare familiar fractions.

Lesson context
Recap on earlier fraction work. Discuss how several unit fractions can combine to give totals such as 2. Using representations of ½ and ¼ circles, engage individuals/pairs in creating different combinations of these for a sum total of two units.

Setting the homework
The task is closed in that there will be just sufficient cut-out pieces to cover the part circles, although this is provisional on these being distributed in a particular way. Explain that some shifting of the pieces between the shapes may be necessary before gluing the pieces down.

Back at school
Show how fractions such as ⅝ result from the addition of ½ and ⅛ by showing that ½ = ¾ = ⅝.

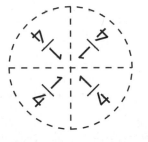
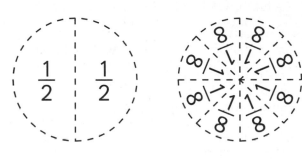

Fraction wall

one whole							
½				½			
¼		¼		¼		¼	
⅛	⅛	⅛	⅛	⅛	⅛	⅛	⅛

p122 FRACTION PIECES (2)
PRACTICE EXERCISE

Learning outcomes
- **Recognise unit fractions and use them to find fractions of shapes.**
- Begin to recognise simple equivalent fractions.
- Compare familiar fractions.
- Estimate a simple fraction.

Lesson context
Draw a large blank rectangle on the board and divide it into four rows. Write 'one whole' in the first row. Divide the next row into two halves and label them. Subdivide the next row into four quarters and the final row into eight eighths. Use this to develop ideas of equivalence. For example: ¾ = ⅝. Erase the wall and provide suitable squared paper for the children to create their own fraction walls and to make up some equivalence sums.

Setting the homework
The task is closed in that there will be just sufficient cut-out pieces to cover the shaded sections, although this is provisional on these being distributed in a particular way. Explain that some shifting of the pieces between the shaded areas may be necessary before gluing the pieces down.

Back at school
Revise complementary fractions with a total of 1 (e.g. ⅝ and ⅜).

p121 WHICH IS BIGGER?
PRACTICE EXERCISE

Learning outcomes
- **Recognise unit fractions and use them to find fractions of shapes.**
- Begin to recognise simple equivalent fractions.
- Compare familiar fractions.

Lesson context
Produce game boards containing two or three unit shapes. Engage the children in a game, using a fraction dice, where they have to roll apprpriate fractions to fill in the shapes on their boards. The final section can only be completed when a dice outcome exactly matches the required fraction.

Setting the homework
Look at the example on the sheet and discuss how it can be derived from the fraction wall. There is no requirement to find all possible statements.

Back at school
Introduce other simple fractions such as ⅕ and ¹⁄₁₀. Test the children's understanding by asking: *Is ½ larger than ¹⁄₁₀?*

p123 FRACTION MATCH
PRACTICE EXERCISE

Learning outcomes
- Begin to recognise simple fractions that are several parts of a whole.
- Begin to recognise simple equivalent fractions.
- Compare familiar fractions.

Lesson context
Use the empty number line approach, this time to extend fractions beyond 1. Demonstrate, for example, counting from 0 to 6 in halves. Provide squared paper to aid the childrens' construction of number lines, and set sequences appropriate to the different ability groupings in your class.

Setting the homework
The homework exercise further extends knowledge of the theme, introducing the decimal equivalent of the most commonly used fractions.

Back at school
Check the work and make links between decimal fractions and fractions of measures such as money and mass.

p124 NUMBER CRUNCH

PRACTICE EXERCISE

Learning outcomes
- Use knowledge that addition can be done in any order to do mental calculations more efficiently. For example: put the larger number first and count on; find pairs totalling 9, 10 or 11; partition into tens and units, then recombine.
- Repeat addition in a different order.

Lesson context
Recap on the idea that addition of numbers can be done in any order. Write the numerals 1–9 on the board and explain that the children's task is to punctuate this chain of numbers with + signs, without changing the order, so as to make a total as near as possible to 100. Demonstrate an attempt (e.g. 12 + 34 + 56 + 7 + 8 + 9 gives a total of 126). Let the children work individually to find the best solution. Afterwards, discuss calculation strategies.

Setting the homework
This exercise further revises the commutative nature of addition.

Back at school
Compare the different strategies.

p125 TAKE THAT

PRACTICE EXERCISE

Learning outcomes
- Use informal pencil and paper methods to support, record or explain HTU – TU, HTU – HTU.

Lesson context
Refer back to earlier examples of finding difference by counting on along a number line. Develop these ideas involving three-digit numbers.

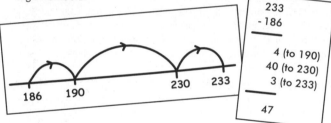

```
        233
      - 186

        4 (to 190)
       40 (to 230)
        3 (to 233)

       47
```

Write some two- and three-digit subtraction problems on the board in a vertical format, using the vertical method to support their mental visualisation of a number line.

Setting the homework
This activity offers further practise of counting on to identify the difference between two numbers.

p126 TIME DIARY

INVESTIGATION

Learning outcome
- Read and begin to write the vocabulary related to time.
- **Use units of time (minute, hour, week, month, year)** and know the relationships between them.
- Use a calendar. Read the time to 5 minutes on an analogue clock and a 12-hour digital clock, and use the notation 9:40.

Lesson context
Review the format/use of a calendar. Ask related questions: *On which day does Christmas Eve fall this year?* Provide a sheet with a set of blank clock faces and some random times to the nearest five minutes. Ask the children to draw the hands on the clocks and to record the times underneath in digital form.

Setting the homework
Talk through the task, reminding children of the conventions for recording time.

Back at school
Share some of the examples. Discuss the use of am/pm and explain why the 24-hour clock is sometimes used.

p127 DRINKS ALL ROUND

INVESTIGATION

Learning outcome
- **Solve a given problem by organising and interpreting numerical data in simple lists, tables and graphs,** for example: pictograms.

Lesson context
Collect data on the class with a theme suitable for presenting as a one-to-one pictogram. (You may like to refer to examples in the NNS *Framework for Teaching Maths*, discussing the data and the form of presentation.) Ask the children to present the data in this form.

Setting the homework
This task needs the support of family and friends and so it may be best to set it over a weekend.

Back at school
Talk about the summarising of data to identify an overall favourite. The notion that the smallest total indicates a favourite may seem somewhat strange to some children.

p128 FAVOURITES

PRACTICE EXERCISE

Learning outcome
- **Solve a given problem by organising and interpreting numerical data in simple lists, tables and graphs,** for example: simple frequency tables.

Lesson context
Conduct a survey of human/vehicular traffic. This should be done by sampling the flow of traffic for a set period such as five minutes, at regular intervals, such as every hour. Customise a chart by, for example, sub-dividing the tally into different types of traffic. Consider different stations for your groups in order to offer comparative data. Redraft the table, adding totals to create a frequency table. Discuss/interpret results.

Setting the homework
This final homework requires a few days for completion. The task is relatively open-ended and may result in some fruitful ideas to pursue as a class.

Back at school
Identify some of the well-judged headings, considering the potential for class investigation.

Name:

Order by phone

- Use a familiar telephone number.

- How many *different* two-digit numbers can you make from this?

The telephone number is _____.

- Now put the numbers in order, starting with the smallest.

- Have you found them all?

Dear Helper,

The actual number selected is not particularly important. What counts is whether numbers can be created and listed in order. If a short telephone number is used, this will help to limit the number of possible combinations. If this proves too easy, think about using a slightly longer telephone number.

Name: _____

Numbers on a plate

P241 XRF

- Choose a car number plate with three numbers.

The number plate is _____ .

- How many *different* three-digit numbers can you make?

- Now write all of these three-digit numbers in full, using words such as 'hundred'.

_____

_____

_____

_____

_____

_____

Dear Helper,

The actual number selected is not particularly important. What counts is whether you child can create the numbers and list them in order. If this proves too easy, think about using a four-digit number. You may need to help with writing the numbers in words. If you have time, write out a 'conversion chart': 10.....ten, 20....twenty, and so on. This will give a lot of support.

Rounding game

You will need: someone to play against and a book with at least 100 pages.

- Take turns to open the book.

- Round that page number to the nearest 10.

- Record your page number next to that multiple.

- Continue taking turns until one person has rounded the opened pages to each of the multiples of 10.

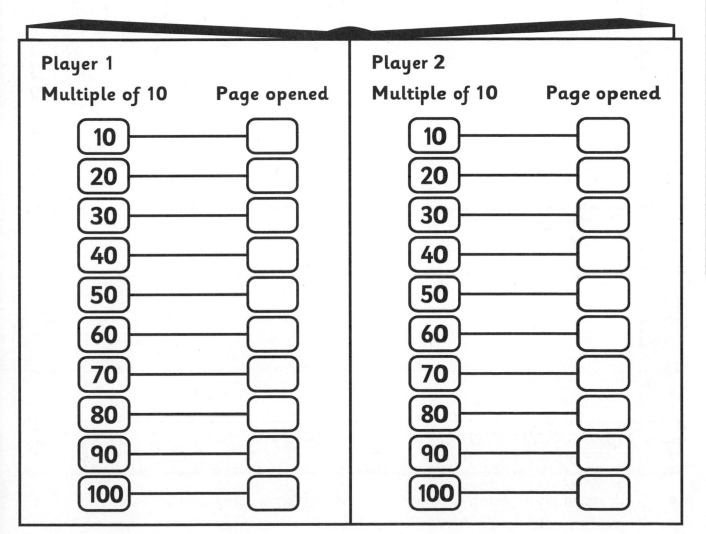

Player 1

Multiple of 10	Page opened
10	
20	
30	
40	
50	
60	
70	
80	
90	
100	

Player 2

Multiple of 10	Page opened
10	
20	
30	
40	
50	
60	
70	
80	
90	
100	

Dear Helper,

Rounding a number up or down to the nearest multiple of 10 is an important skill to help with estimating and checking the results of calculations. If the number ends in a digit less than 5, the number is rounded down. For example, 18**3** is rounded down to 180, since 183 is closer to 180 than 190. If the number ends in a digit of more than 5, that number should be rounded up. 12**6** is rounded up to 130, since 126 is closer to 130 than 120. If the number ends with the digit 5, the convention is always to round up rather than down. So 8**5** is rounded up to 90.

Name:

It all adds up

- Look at these four numbers and see how many different ways they can be combined. The first one has been done for you.

$12 + 7 + 8 + 3 = 30$

- How many *different* ways are there?

- Look carefully at each arrangement. Which way is the easiest for you to calculate? Why is it the easiest?

Dear Helper,

It is important for children to recognise that we can add up a set of numbers in any order, as the total will always be the same. Talk about the different ways of listing the numbers as a sum and then, looking at the problem from left to right, see if any seem to be quicker to combine. Often it is easier to add up totals when there are convenient pairings, for example, a 6 and a 4 makes 10.

Name:

Target 20

Use a pack of 0–9 digit cards, shuffled
and placed face down as two rows of
five cards.

- Flip over three cards and find their total.
  If the total is 20, write the combination down.

- Continue playing, turning the cards back over after each attempt.

- See if you can find different ways of making 20.
  After a while you will begin to learn where the numbers lie.

- Work with your helper, taking turns to play the game.

These combinations of three cards make 20:

Dear Helper,

Memory games such as this are a useful way of sharpening number skills. As you play this game, try to discuss the combinations to help you get inside the addition strategies. Similar activities can be devised using packs of playing cards: *Can you find four cards to give a total of 25?* Another game is to take turns to turn cards over from the top of the pack. You must keep a running total without going over 21. If you do, those cards are given to your opponent!

Name:

Make it pay (1)

You will need: a set of 0–9 number
cards and these coins: 10p, 5p, 2p, 2p, 1p.

- Pick two or three cards and calculate
  the total. Do not go beyond 20.

- Now use some of your coins to make
  that amount in pence.

- Show the coins and the totals in the
  space below.

- Continue on the back of the sheet if
  you need more space.

Dear Helper,

Given the importance of money in our lives, it is vital that we know how to use it in everyday situations.
The strategy for making a total such as 17p, by starting with the largest possible coin, needs to be
reinforced. The activity can be more challenging by working with larger totals and a wider range of coins.

Number sentences

- Use the numbers given opposite to make as many addition and subtraction sentences as you can:

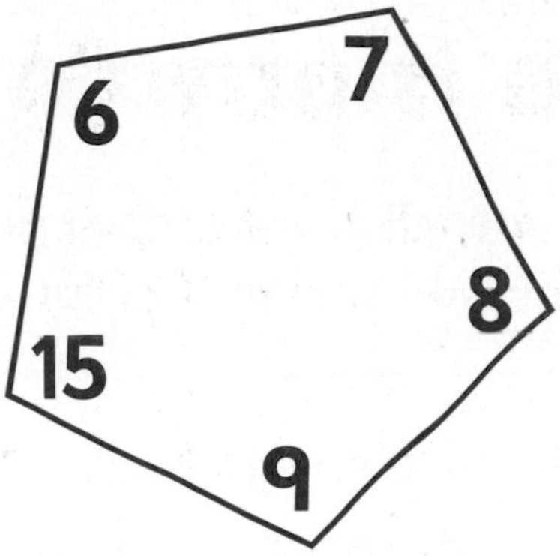

6   7

8

15

9

I can make these sentences:

Addition	Subtraction

Dear Helper,

We often tend to think of addition and subtraction as two completely separate operations, when actually they are closely linked. When finding the **difference** between 13 and 20, for example, children often find it easiest to count on from 13 to 20, a process more akin to **addition**. If the task is completed quickly or with ease, make up some number sentences using a set of larger numbers such as 12, 16, 15, 13 and 28.

Make that total

- Make the totals on these cards using the numbers 4, 5 and 7 only. You can use the same number more than once.

13

16

20

- Now make these totals using 4, 5 and 7 **every** time.

21

23

25

27

28

Dear Helper,

This activity features an element of problem solving, where the correct solution is not immediately obvious. Children sometimes need reassurance that it is acceptable to experiment with numbers and, inevitably, to get it wrong sometimes. Support at this stage of 'trial and improvement' can do a lot to boost a child's confidence. You may also notice that the skill of repeatedly adding the same number, in order to 'approach' the target number, is a feature of this task. This should help, therefore, to support knowledge of multiplication facts.

Make 20

You will need: a set of number cards from 0 to 20.

- Arrange these cards, face down, as three rows of seven.

- Try to play this game with someone else.

- Take turns to try to flip two cards with a total of 20.

- If the total is 20, you collect the cards. If not, they must be turned back over.

- The winner is the person with the most cards when all have been collected.

- Write down all the combinations you found in the space below.

Ways of making 20:

Dear Helper,

The aim of this task is to rehearse adding two numbers with a total of 20 and, at the same time, to look at subtraction *from* 20. Try to take opportunities to talk about the work as you play the game. In particular, use supportive comments such as: *You've turned a 6, so you need the number that is 20 less 6.* Eventually, the aim is for all children to develop instant recall of addition and subtraction facts from and to 20 but, until that time, it is perfectly acceptable to use fingers, number lines or whatever helps.

Name:

What's in a name?

• Cut and paste these 12 items into four 'families'.

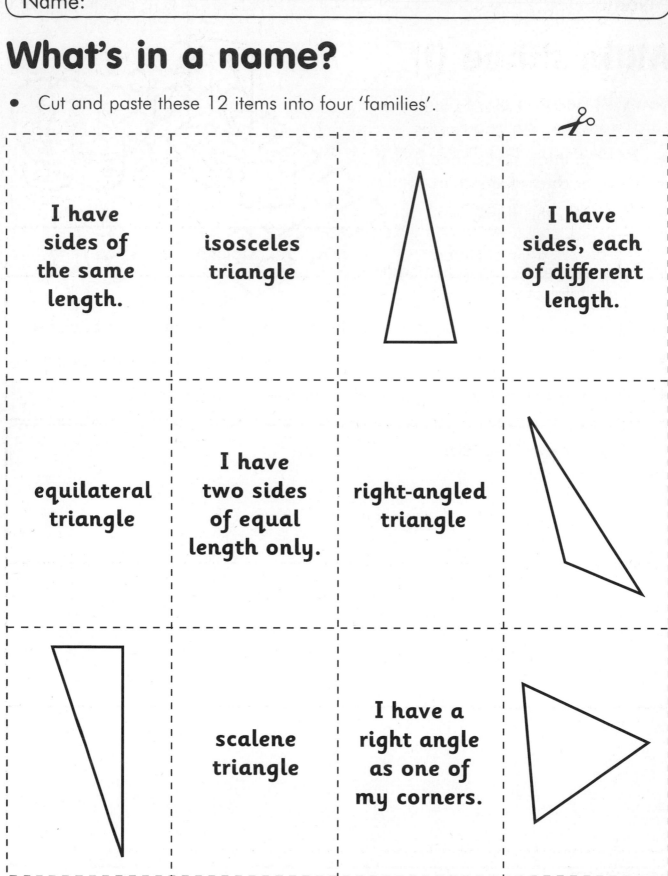

I have sides of the same length.

isosceles triangle

I have sides, each of different length.

equilateral triangle

I have two sides of equal length only.

right-angled triangle

scalene triangle

I have a right angle as one of my corners.

Dear Helper,

It is particularly important for children to know some of the technical terms used to describe shapes. This exercise builds's on ideas explored in lesson time. Some dictionaries (electronic and book-based) feature some of this vocabulary, but it might be possible for your child to recall this from memory. Encourage the idea of 'doing your best', possibly starting with the cards that you know go together.

100 MATHS HOMEWORK ACTIVITIES • YEAR 3 TERM 1

Name:

All in shape (1)

- Find some 3-D shapes at home.
  The kitchen is a good place to start
  (for example, packets and tins of food).

- Draw each shape and label it with
  its mathematical name.

Picture of shape	Mathematical name

Dear Helper,

This activity deals with the topic of solid shape, sometimes referred to as 'three-dimensional geometry'.
It may be quite challenging to find a wide range of shape types, as most containers and packages tend to
fall into relatively few mathematical categories. If possible, talk about the properties of a given shape.
For example, if the shape is a cube, what exactly makes it so? You might talk about the surfaces: *Are they
flat/curved/how many are there?* Or talk about the number of edges and vertices ('corners' is a less
precise, and potentially misleading, term).

Name:

All in shape (2)

- Look at the faces of some objects at home.

- See if you can find some different 2-D shapes.

- Draw each shape and label it with its mathematical name.

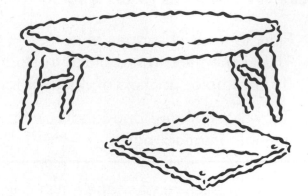

Picture of shape	Mathematical name

Dear Helper,

Looking at 2-D shapes around the home involves looking at the faces of real objects. In theory at least, 2-D objects do not exist in the real world because they have no thickness. Flat (sometimes called plane) shapes can be either regular or irregular. A regular shape, such as a square, has all its sides the same length. All the internal angles measure the same too. Fabric and wallpaper patterns can be a rich source for geometric shapes.

Name:

Look at the label

- Look at food packages and tins to find out the mass of each item.

- Put them in order of mass.

Item	Mass

- Draw each item, starting with the lightest.

Dear Helper,

Although we now use kilograms extensively, you can still find imperial measures (such as pounds and ounces) in evidence. It is important for children to have a working knowledge of mass, and be able to estimate how much something weighs. For your information, the metric unit of a kilogram is approximately equal to 2.2 pounds, and 1000 grams are equal to a kilogram (1000g = 1kg).

How long?

- Find things that you would use to measure length.

- Draw a picture of these and measure something with each one.

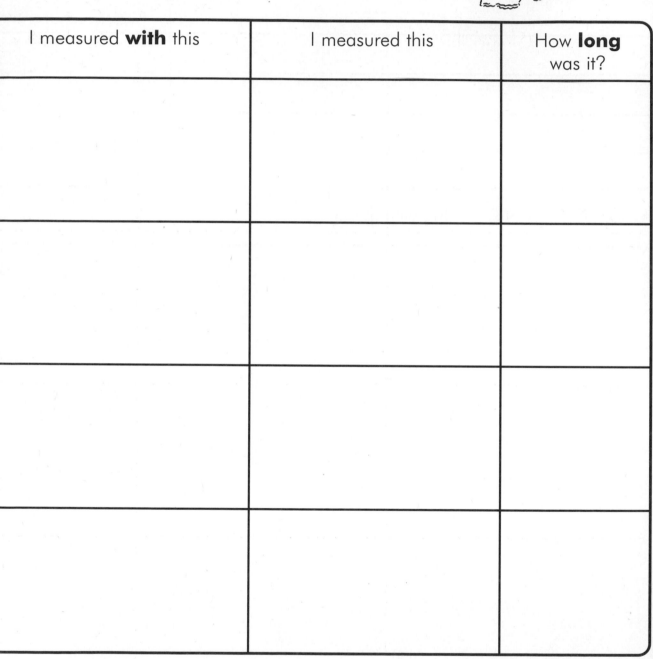

I measured **with** this	I measured this	How **long** was it?

Dear Helper,

Although children need to have a working knowledge of feet and inches, the preferred units are now metres and centimetres. They need to be able to make sensible estimates of length and be able to record lengths appropriately. Your child needs to be able to convert between centimetres and metres, so a length such as 128cm can also be written as 1 metre 28 centimetres or 1.28m. Later on children must also be familiar with the use of millimetres (1280mm in this case). Hopefully, you may have some measuring devices relating to DIY, schoolwork or hobbies.

Name:

More measures

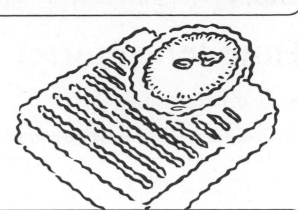

- Find equipment that you would use to measure mass.

- Draw a picture of these and find some things to weigh.

I measured **with** this	I measured this	How **heavy** was it?

Dear Helper,

In school, we tend to talk about the 'mass' of items rather than their 'weight'. Although we use 'weight' in everyday language, the use of 'mass' is technically more precise. You will still be 'weighing', but try to refer each item's 'weight' as its 'mass'. Hopefully, you may have kitchen scales and/or bathroom scales available for this activity.

PHOTOCOPIABLE

43

Name:

Longer – shorter

- Use a flexible tape.
- Find some items **longer** and **shorter** than 1 metre.
- Draw them in the correct column.
- Write the length next to each picture.

Longer than 1 metre	**Shorter** than 1 metre

Dear Helper,

The metre is now our standard metric measure of length, so it is important for children to be familiar with this length. When recording lengths in metres, it is usual to write in decimal format, so 136cm is written 1.36m.

Name:

How many?

- Think of a large collection of things you have at home.

- Count out the first ten and use this to estimate the number of items in your collection.

- Now group your collection in tens.

I have counted a collection of:	I think I have this many items in my collection:
How many groups of ten?	Did you have any left over? How many?

How many items **altogether?**

Dear Helper,

Estimation is a valuable skill, which can be developed. Children need to be encouraged to make sensible estimates. An estimate is not a guess. Estimating the number of people watching a movie, for example, can be estimated by counting ten people and trying to judge how many similar-sized groups are in the audience. This strategy could usefully be applied to this homework.

Name:

Sequences (1)

- Shuffle a pack of 0–9 cards and place them face down.

- Pick a pair of cards. Make the smallest two-digit number you can with them.

- Now count back in tens until you get back to one of your starting digits.

- Write some number chains down.

For example:

7 and 3 | 37 | 37 — 27 — 17 — 7

- You could try counting on and back in hundreds using three cards each time.

Dear Helper,

Your child needs to be confident in adding on or taking away a number from a given start number. This work develops early aspects of algebra, where each sequence is governed by a rule. Refer to the various digits in terms of their being worth so many ones (or units), so many tens and so many hundreds. In 234, for example, the middle digit is actually worth three 10s, i.e., 30.

Name:

Pigs and ducks

- Draw pictures of animals with a total leg count of 16.

- You are only allowed to draw ducks, pigs or combinations of pigs and ducks.

Dear Helper,

This problem requires a thorough approach to try and find all the possible answers to this question. It is often helpful to think about a set number of one animal, say, pigs, and then to see how many ducks would be needed to 'make up' to the given total. To complete this activity, the puzzler must be satisfied that all possible answers should have been found. This exercise also helps children to count in groups of 2 and 4, and to know that total rapidly. How quickly, for example, can your child tell you how many legs six pigs would have?

Name:

Target 80

- Use the numbers 1, 2, 3, 4, 5 and 6 in that order.

- Insert some addition signs to try to make a number near to 80.

Can you do better?

	total
1 + 2 + 3 + 45 + 56	107
1 + 23 + 4 + 5 + 6	39

Dear Helper,

Sometimes mathematics requires the use of a method called 'trial and improvement'. To do this, your child will need to 'stick at it' and 'have a go'! Children need encouragement and praise, sometimes improving on earlier trial combinations to get closer to the target. You might also find it helpful to advise on some addition strategies. In the second example above, for instance, some people begin with addition of 4 and 6 to give 10.

Name:

Superbugs!

Your superbug just keeps multiplying!

- Add an extra identical bug after each row and count the total number of legs.

	Number of legs

Try to learn the number sequence.

Dear Helper,

As well as learning the multiplication facts, it is important that children fully understand how these are developed. They can then use these strategies for finding the answer when memory lets them down. Multiplication includes the aspect of repeated addition and this is what lies behind this activity. You can help further by talking about the sequence of answers. You could also ask questions out of sequence such as: *How many legs on six bugs? Three bugs?* and so on.

Name:

Supershapes!

These shapes grow on you!

- Add an extra (identical) shape on each row and count the total number of corners.

	Corners

Try to learn the number sequence.

Dear Helper,

This exercise gives further practice in learning the multiplication facts. Once finished, you should help your child to recall the sequence in order and then ask questions out of sequence. If an answer is not known, try to encourage the finding of a known fact that is near to it, and then adjust up or down by adding on or taking away.

Name:

Fair shares

- Find a collection of objects to share. Select no more than 30.

- Try to share it equally between 2, then 3, then 4.

- Put a tick if it shares equally, but put a cross if it does not share equally.

- Try different numbers in your collection.

Number in collection	Does it share between 2?	Does it share between 3?	Does it share between 4?

- Can you find a number giving three ticks?

Dear Helper,

This activity develops the idea of sharing as a method of division. Children need to be familiar with the idea of sharing with someone else: 'One for me and one for you, one for me and one for you'. If possible, involve others so that sharing between 2, 3 and 4 can be seen as a 'real activity'. When the sharing is complete, talk about whether everyone has an equal share or whether there are any remainders.

PHOTOCOPIABLE

Making money

- Use real coins to help you.

- Find an amount in **two** *different* ways.
  For example, for 5p, you could use:
  1p, 1p, 1p, 1p, 1p or 2p, 2p, 1p.

- You can draw the coins or write them
  as numbers.

Amount	or

- Try to do the same again, this time with larger amounts.

Amount	or

Dear Helper,

Your child needs to be confident in using the full range of coins and notes. For this activity, it will be helpful for you to provide as wide a range of coins as possible. Start with amounts no greater than 50p and then work towards £1.00. Try to work with amounts that offer a balance of challenge and 'manageability' for your child. You might also need to check that the amounts selected are recorded in the appropriate manner. 50 pence, for example, should be written in pounds as £0.50, not £0.50p.

Name:

What shall I buy?

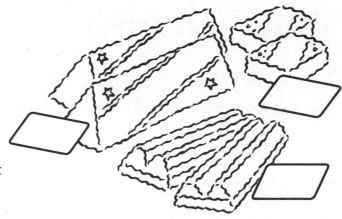

- You can buy any two items from those shown.

- Draw all the different combinations and the coins you would use to buy them. You are allowed to buy two of the same thing.

- Use real coins to help you.

Dear Helper,

This task develops three different skills, which should provide a good basis for discussion. First, there is the important skill of finding as many different pairings as possible, something that is best done by being systematic. Second, there is a requirement to add two quantities accurately, helped by using real coins. Finally, there are decisions to be made about which coins to give for payment. There are no right answers about the coins, although it might be worth considering the fewest coins each time.

Name:

Shape share

- Fill each shape with three colours.

- Colour half ($\frac{1}{2}$) in yellow, a quarter ($\frac{1}{4}$) in blue and a quarter ($\frac{1}{4}$) in red.

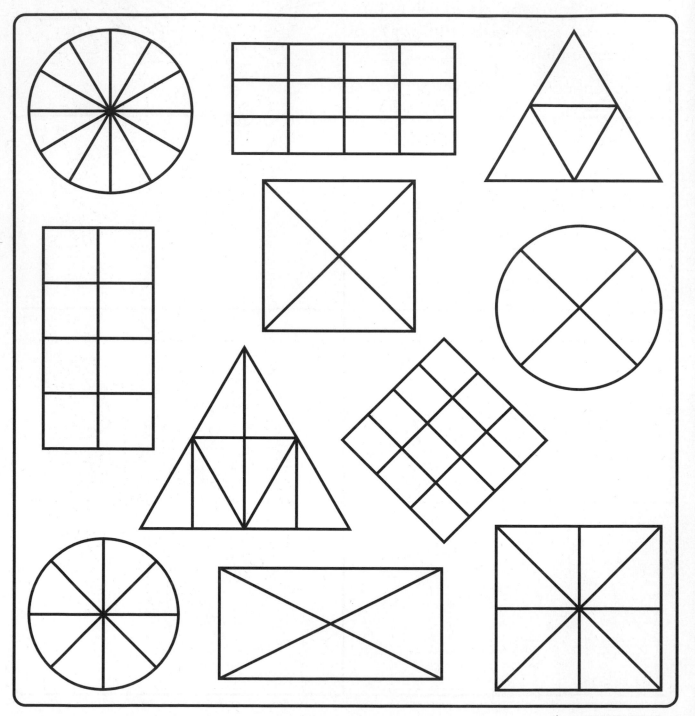

Dear Helper,

It is important that children should recognise that, for example, a quarter of an object or shape is more than simply '1 of 4' pieces, as each quarter needs to be equal in size. Children need many examples to fully appreciate how, for example, two quarters make a half. The use of different numbers of sub-divisions also helps to link the idea that a quarter of 12 (sections) is 3 (sections).

Name:

Dividing out

- Colour half ($\frac{1}{2}$) of each set in red and
  a quarter ($\frac{1}{4}$) of each set in blue.

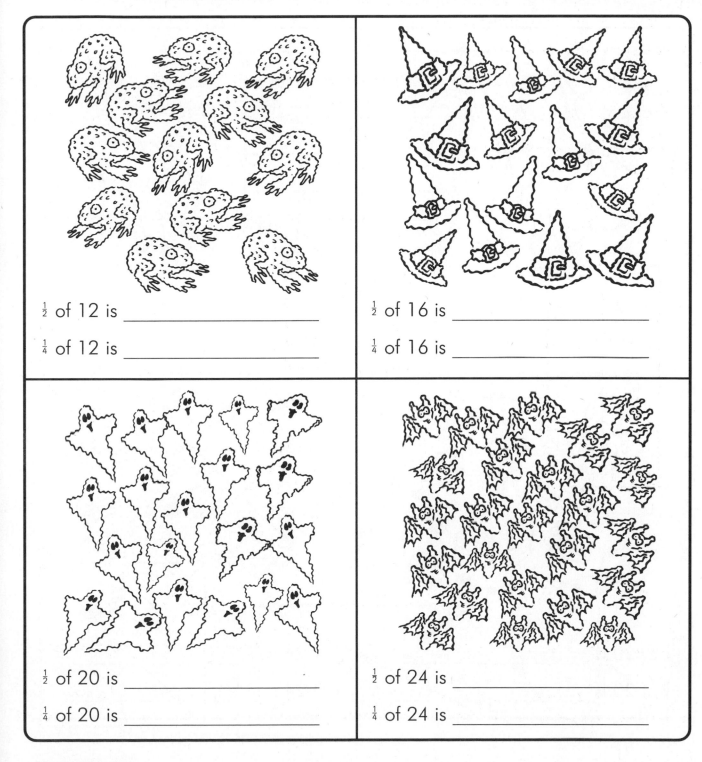

$\frac{1}{2}$ of 12 is _____

$\frac{1}{4}$ of 12 is _____

$\frac{1}{2}$ of 16 is _____

$\frac{1}{4}$ of 16 is _____

$\frac{1}{2}$ of 20 is _____

$\frac{1}{4}$ of 20 is _____

$\frac{1}{2}$ of 24 is _____

$\frac{1}{4}$ of 24 is _____

Dear Helper,

This exercise focuses on fractions of a quantity. Sometimes children benefit from having similar sets of counters, rather than relying on the drawings provided above. Using real objects can then make use of sharing into two, then four, equal piles. Checking that each pile contains the same number of objects confirms that equal fractions have been found.

Name:

Part shares

- Colour a quarter (¼) of each set in
  blue and three-quarters (¾) in red.

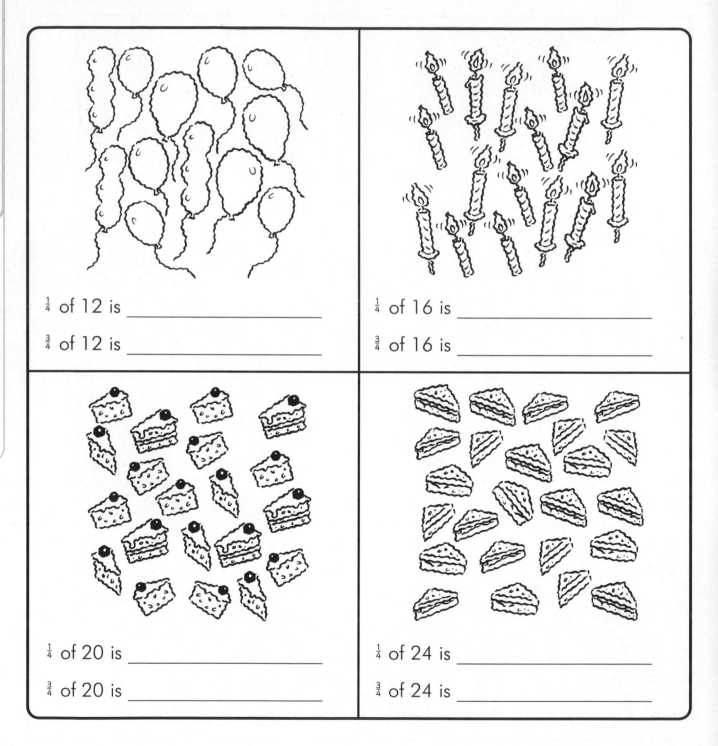

¼ of 12 is _____

¾ of 12 is _____

¼ of 16 is _____

¾ of 16 is _____

¼ of 20 is _____

¾ of 20 is _____

¼ of 24 is _____

¾ of 24 is _____

Dear Helper,

This activity requires children to find three-quarters of a quantity. There are two helpful strategies. The first involves finding one quarter and 'multiplying up' three times to find three-quarters. A second strategy relies on working out half and then a quarter of the quantity, and adding these two answers together. Some real objects, such as counters, for sharing out will help.

Jumps

- Use the number square to help you add.

1	2	3	**4**	5	6	7	8	9	10
11	12	13	**14**	15	16	17	**18**	19	20
21	22	23	24	25	26	27	28	29	30
31	**32**	33	34	35	**36**	37	38	39	40
41	42	43	44	45	46	47	48	49	50
51	52	**53**	54	55	56	**57**	58	59	60
61	62	63	64	65	66	67	68	69	70
71	72	73	74	**75**	76	77	78	79	80
81	82	83	84	85	86	87	88	89	90
91	92	93	94	95	96	97	98	99	100

- Add 22 to each of the bold numbers. Colour these in red.
- Add 18 to each of the bold numbers. Colour these in yellow.
- Add 24 to each of the bold numbers. Colour these in blue.

Dear Helper,

The use of the number grid has proved to be a useful approach in moving children on to mental methods of calculation. Children are taught that moving along a row means counting on in ones, while moving down a column involves adding on in tens. Adding 22, for example, can then be seen as moving along a row two places and then down that column two places (or vice versa). Children can become mentally agile at manipulating numbers. Adding on 59, for example, can be thought of as 'add 60, less 1'.

Name:

Giant jumps

- Bridge through the nearest multiples of 10 to add these numbers. The first one has been done for you.

add 26 to 16

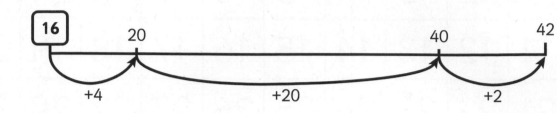

add 35 to 17

add 44 to 29

add 53 to 38

- Now make one or two of your own.

Dear Helper,

These questions all use an 'empty number line'. This approach is useful for many children as they move into more formal ways of recording. It also provides another useful visual 'prop' for times when questions need to be worked out 'in the head'. If your child finds this work relatively straightforward encourage them to make up some of their own questions using numbers beyond 100.

100 MATHS HOMEWORK ACTIVITIES • YEAR 3 TERM 1

Name:

About time

• Draw pictures and complete these clocks to create
  a time diary of things you do at the weekend.

Dear Helper,

Children come to an understanding of time at widely different rates. Regular exposure to telling the time is a useful strategy. The presence of traditional clock faces as well as digital timers helps too. Children often need help in reading clock-face and digital times. Traditional clocks deal with time to and past the hour, while digital displays only consider time past the hour (for example, twenty to four as opposed to 3.40). In school, children will be taught other conventions, such as the use of am and pm and the 24-hour clock.

Name:

TV (1)

- Look through a TV listing for this week to make lists of your favourite programmes.

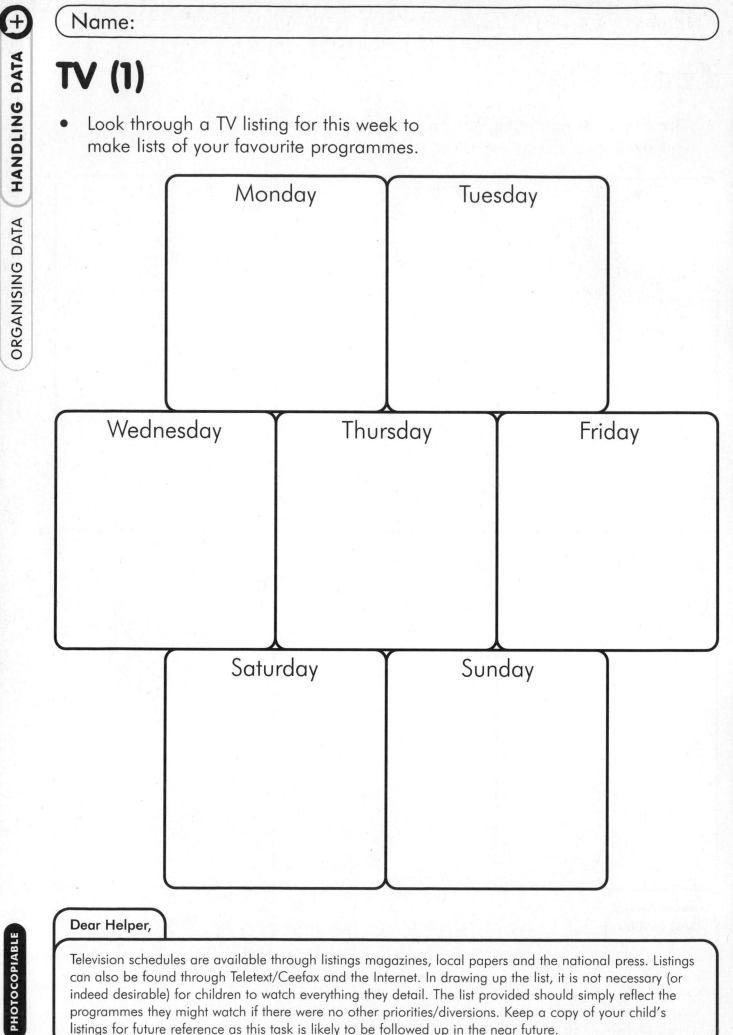

Monday

Tuesday

Wednesday

Thursday

Friday

Saturday

Sunday

Dear Helper,

Television schedules are available through listings magazines, local papers and the national press. Listings can also be found through Teletext/Ceefax and the Internet. In drawing up the list, it is not necessary (or indeed desirable) for children to watch everything they detail. The list provided should simply reflect the programmes they might watch if there were no other priorities/diversions. Keep a copy of your child's listings for future reference as this task is likely to be followed up in the near future.

Name:

TV (2)

- Use the TV programmes you listed on the last homework sheet to write down and total the length of viewing on each day.

| Monday | Tuesday |

| Wednesday | Thursday | Friday |

| Saturday | Sunday |

Dear Helper,

Use the listings from the earlier homework. Your child may need help in working out the length of each programme. You may find it helpful to use a clock face to work out the difference in minutes between the start of the programme and the beginning of the next. Help will also be required in finding the total time required to watch the programmes, although it must be emphasised that this is a 'wish list' and not the actual time to be spent.

Name:

TV (3)

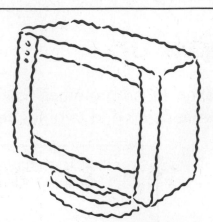

- Survey the types of programme others like to watch.

- They can only choose from this list and they must rank them in order from 1st to 5th.

Name	Cartoons	Films	Drama	Soaps	Comedy

Dear Helper,

Your child will need the support of friends and family to complete this task. Some help may be also be needed in giving examples of each programme type, in order to focus the mind on what it is they are voting for. When the table is complete, a useful discussion might concern whether an overall 'winner' can be established by collating the results from each category.

Going up

• Crack the code! Cut out these numbers and glue them in order into the spaces provided below, starting with the smallest number.

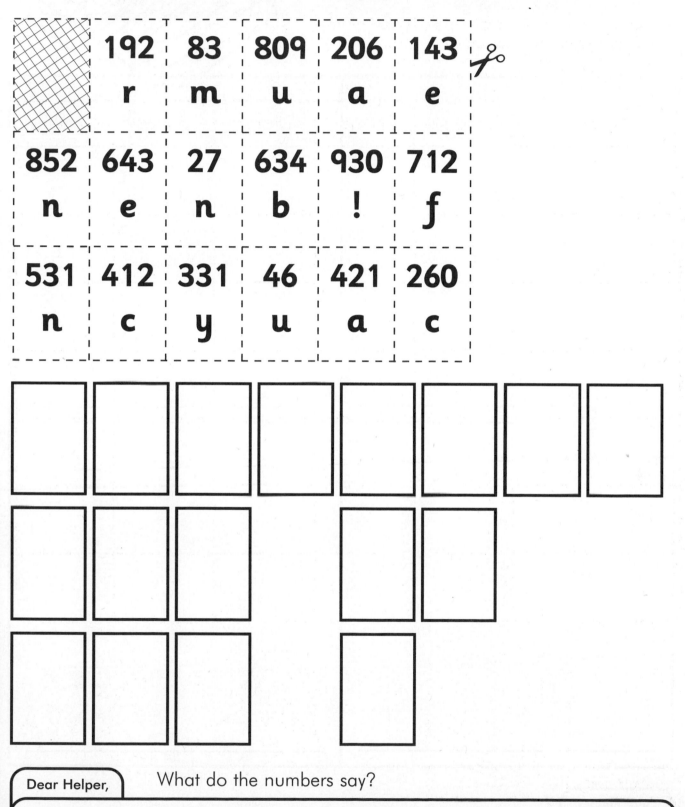

	192	83	809	206	143
	r	m	u	a	e
852	643	27	634	930	712
n	e	n	b	!	f
531	412	331	46	421	260
n	c	y	u	a	c

Dear Helper,   What do the numbers say?

The clever thing about our number system is that any number can be created using the digits 0–9. Being able to order numbers, however, requires an understanding about the value of digits according to where they lie. In 42, for example, the 4 has a greater value than the 4 in 64. When children are confident about putting numbers below 1000 in order, larger numbers can be explored.

Name:

Scales

- Find things in and around the home which have scales for measuring things.

- Draw one in each box below and write what each measures.

Dear Helper,

Your child will need guidance in carrying out this task safely. Drawing a dial from a car dashboard, for example, should only be done under supervision. Help may be needed in showing some of the details on the scales, and deciding what they actually measure.

Name:

Make it up

- Write some number stories using the words and numbers below:

  For example: The sum of 15 and 23 is 38.

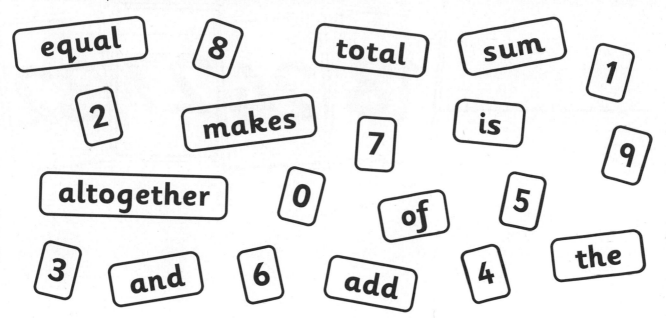

equal   8   total   sum   1

2   makes   7   is   9

altogether   0   of   5

3   and   6   add   4   the

- Have you used *all* the words?

Dear Helper,

Some of the complexity of mathematics comes from the vast range of words we use when talking about the subject. Part of the reason for this is that mathematics should involve clarity and precision, and this requires specialist vocabulary. This activity uses a few words connected with addition, and it would be helpful for your child to read their written answers back to you.

Name:

Number pairs

- How many different addition sums can you complete using any two of these numbers each time?

- Have you found all the possible sums?

Dear Helper,

This activity gives your child some further practice in addition. It is also important, if time permits, to encourage them to try to find all the possible combinations. The sums could be re-written to give totals in ascending numerical order, but this is not essential.

Name:

Make it pay (2)

You will need: a set of numeral cards (from 11–19), and a pile of coins including:

- Pick two or three cards and calculate the total. Do not go beyond 50.

- Now use some of your coins to make that amount in pence.

- Show the coins and the totals in the space below.

Dear Helper,

It is important for all children to be able to add up accurately without the use of pencil and paper. This exercise gives practice in adding up money and choosing an appropriate set of coins to match that value. While there are often several ways to make any given total, your child should be moving towards the use of as few coins as possible by selecting the largest coin denominations possible. It is worth noting that the number of coins required for any total to 50p rarely needs to exceed five.

Name:

Making more money

£1	10p	£2	20p	5p	2p
5p	2p	£1	50p	1p	£2
2p	50p	5p	£2	10p	2p

- Cut out the coin tiles above.

- Make each of these amounts using the cut-out coin tiles.

- For each amount you should use the **least** number of coins.

Amount	Coins used
£1.23	
£3.55	
£2.69	
£2.17	

- Glue the tiles down when you have used all the coins.

Dear Helper,

This exercise gives further practice in choosing suitable coins to pay for things using the full range of coins. There are only just sufficient coins to complete all the questions, so arrange them loosely on the page before gluing them down permanently.

Name:

One stop party shop!

- Work out the cost for each person at a party.
  It may help to pair some items to give you multiples of
  10p and £1.00.

- How much will it cost for a party of 12 people?

Dear Helper,

Your child may benefit from help in this task because it contains several stages and involves relatively large quantities. There are different ways of approaching this task too, so a discussion about how to get started could be particularly useful. It is important to note that some of the costs can be 'paired up' to give more manageable sub-totals which can then be added together.

Name:

Word problems!

- Read each problem and then write your answer and working alongside.

	Answer	Show your working
A young girl is a quarter of her mum's age. The girl is 8. How old is her mum?		
Sarah is 26cm taller than Amir. Sarah is 112cm tall. How tall is Amir?		
I have some money. My friend has twice as much. If we have 24p altogether, how much do we have each?		
I divide a cake into eight pieces. Each piece weighs 120g. How much does the whole cake weigh?		
Find two odd numbers which multiply together to give 63.		

Dear Helper,

In mathematics there is a tendency to worry only about getting the 'right' answer. In this activity it is important for your child to show their working out, even if the method may seem unduly long-winded. This helps your child's teacher to see what needs to be taught next. Word problems can have extra difficulties if some of the mathematical terms are not understood. Children are sometimes unclear about which operation (+, −, × or ÷) to use.

Name:

Shapes in words

You will need: eight straws of two different lengths.

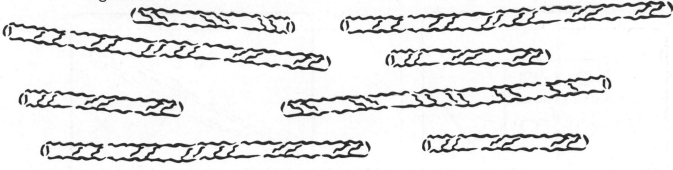

- Use the straws to make and draw the following shapes.
- Write in words what properties give these shapes their names. One has been done for you.

Shape	Drawing	Reason
Square		All the sides are the same length. The shape has four right angles.
Rectangle		
Equilateral triangle		
Parallelogram		
Rhombus		

- Can you make a kite shape?
- Can you make a different type of triangle?

Dear Helper

As well as being able to recognise shapes, it is also important for children to know what qualities, called 'properties' in mathematics, give shapes their names. To this end, it is necessary to describe shapes precisely. In describing a square, for example, it is not enough simply to say that it has 'four equal sides'. Some dictionaries, fact-finder books or encyclopaedias offer useful guidance on shape definitions, if further help is needed.

Name:

Symmetry

The outline of this object has
'line symmetry':

This does not:

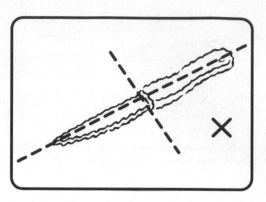

- Use things at home to draw shapes with and without line symmetry.

These have line symmetry	These do not have line symmetry

Dear Helper,

'Line symmetry' is all around us, in both natural and manufactured products. For an object to have line symmetry, you must be able to draw a line in such a way that the two halves match. Strictly speaking, we are only dealing with two-dimensional (2-D) shapes so, if you are dealing with a solid object, it is the faces that should be the focus of attention. Drawing a line through the face of a dice is one such example.

100 MATHS HOMEWORK ACTIVITIES • YEAR 3 TERM 2

Name:

About time

- Find out about forthcoming birthdays for people you know.

- Use a diary or calendar to find out the days on which these birthdays fall.

Name	Birthday (date)	Day of week	Number of days in that month

- Work out which of these people have their birthday first.

- How long is it until their birthday?

Dear Helper,

Although the *date* of our birthday never changes, the peculiarities of our calendar system mean that our birth *day* can vary from year to year. A current diary or calendar will give details of the days and dates for this current year. Your child should also be aware that the length of each month varies too. It is useful to know those months with fewer than 31 days.

Picture this

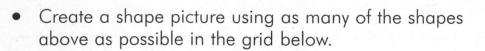

- Create a shape picture using as many of the shapes above as possible in the grid below.

- Colour the individual shapes in a range of colours, one colour for each type of shape.

- Now colour your shape picture so that the shapes are matched by colour.

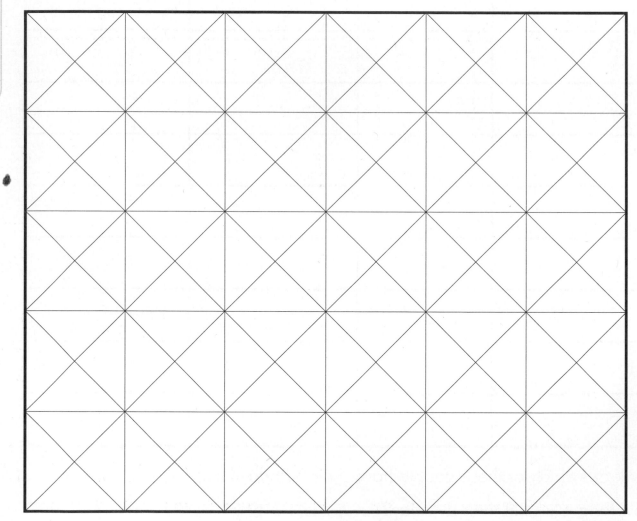

Dear Helper,

This activity requires your child to draw shapes with a ruler, to colour-code shapes of the same type and to know the names of each shape in order to talk about their picture. Discuss the names of the shapes with your child. Try to encourage simple drawings made up from large copies of the shapes.

Name:

Boxes

The pictures below show all the different ways of stacking three cube boxes against a wall.

Each box must share one of its faces with another:

These ways are not allowed:

✗ gaps   ✗ overlaps

- See if you can find all the different ways of stacking four boxes in the way allowed.

- Have you found them all?

Dear Helper,

This activity could be made a little more real if you have four toy bricks, stock cubes or dice, for example. The successful problem solver is the person who works in an organised and methodical way to find all the possible solutions. The task will be discussed at school, but it is worth knowing that the answer lies between 5 and 10.

Name:

On the grid

- Look at the coordinates of the square. Double the coordinates to complete the table below.

Co-ordinates of shape	(1,2)	(2,4)	(4,3)	(3,1)
Double				

- Use these new coordinates to draw the enlarged square.

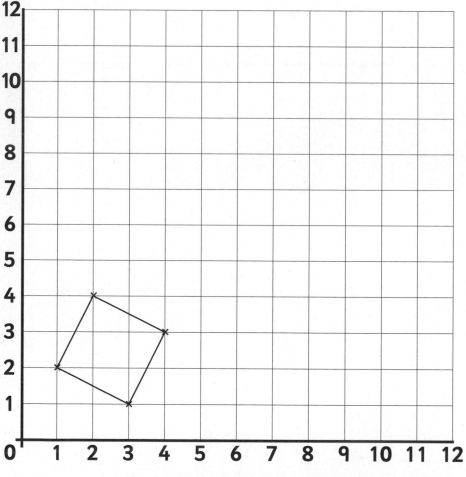

- Can you add a third square by trebling the coordinates?

- If your finished work looks like a pattern you could colour in the different shapes and overlaps.

Dear Helper,

When plotting coordinates on a grid, it is important that children match the first number of the pair with the numbers on the base of the grid. The second number is matched with the vertical. On these grids, the coordinates always lie on the intersection (crossing point) of a horizontal and vertical line. Coordinates are used frequently in both mathematics and geography, and children may see these in road atlases and on maps.

Name:

Getting into shape

- Use this special isometric paper to draw these shapes:

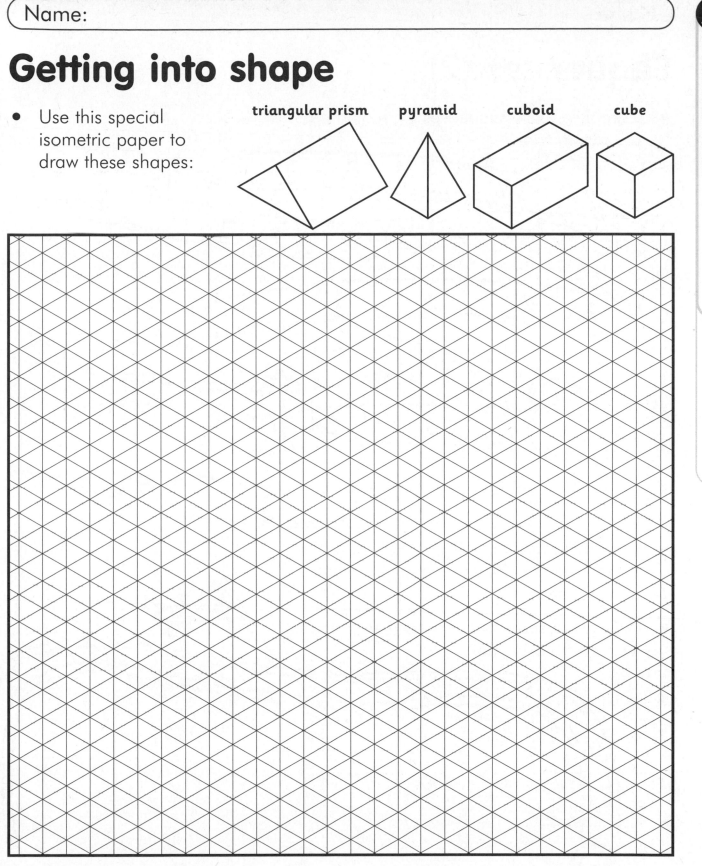

triangular prism   **pyramid**   **cuboid**   **cube**

- Remind yourself about the number of faces, edges and vertices in each shape.

Dear Helper,

Drawing representations of solid objects on paper can be quite a problem for children and adults! This activity includes a grid designed to help construct these pictures. The words used to describe a shape are very specific: the surfaces are known as 'faces', the corners are 'vertices' and the edges are known simply as 'edges'!

Name:

Sequences (2)

- Complete these sequences by adding on or back in twos.

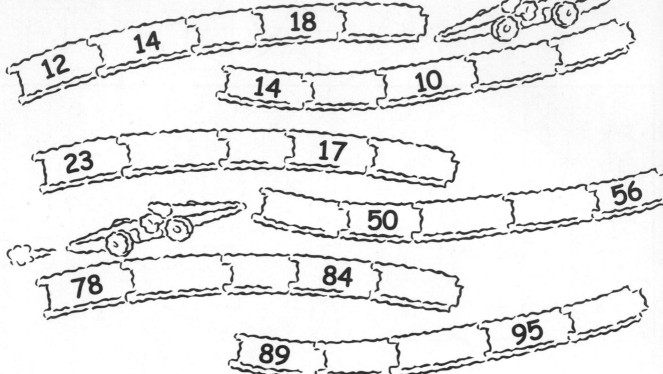

12 | 14 | | 18 | |

| 14 | | 10 | |

| 23 | | | 17 | |

| | 50 | | 56 |

| 78 | | 84 | |

| 89 | | | 95 | |

- Colour the odd numbers in one colour and the even numbers in another. What do you notice?

- Now make sequences of your own using two- or three-digit numbers.

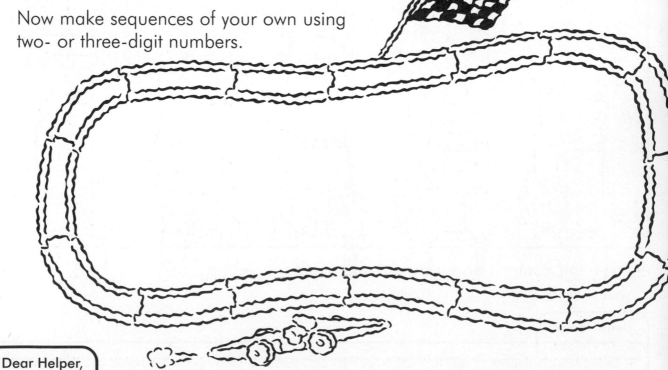

Dear Helper,

This activity involves patterns with sequences which increase and decrease. Your child may need support to ensure that the missing numbers are chosen correctly so that the sequence changes in equal-sized steps.

100 MATHS HOMEWORK ACTIVITIES • YEAR 3 TERM 2

Name:

Sequences (3)

- Complete these sequences by counting on or back.

- Colour the odd numbers in one colour and the even numbers in another.

- Are there any patterns?

- Now make sequences of your own using two- or three-digit numbers. You can count on or back in steps of 3, 4 or 5.

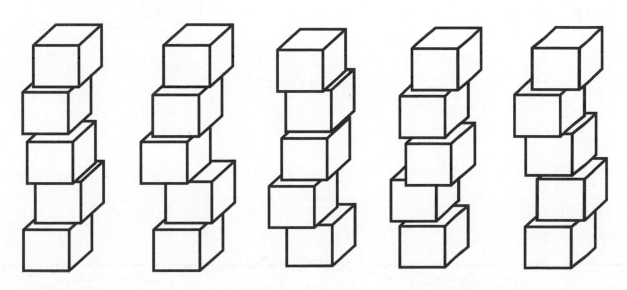

Dear Helper,

This homework offers further practice of sequences. It is helpful to discuss the different colour patterns created, and the reasons for those patterns. Why, for example, does a sequence alternate in colour when 3 is added each time? Work of this type continues to prepare the groundwork for later study in algebra.

Name:

Statements

- Make some different statements using × and ÷ and = and the numbers opposite.

- One has been done for you.

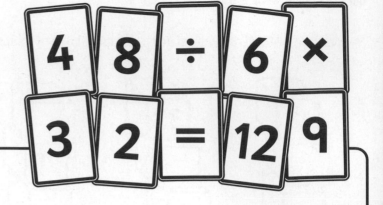

$$4 \times 2 = 8$$

- How many can you find?

Dear Helper,

In much the same way that addition and subtraction are linked, so too are the operations of multiplication and division. Your child should know that, for example, 2 × 8 = 16 is related to the fact that 16 ÷ 2 = 8. Recognising that the two operations are linked will help your child to speed up in their number work.

100 MATHS HOMEWORK ACTIVITIES • YEAR 3 TERM 2

Name:

Take back

- Take steps back to **0** to fill in the spider's web.

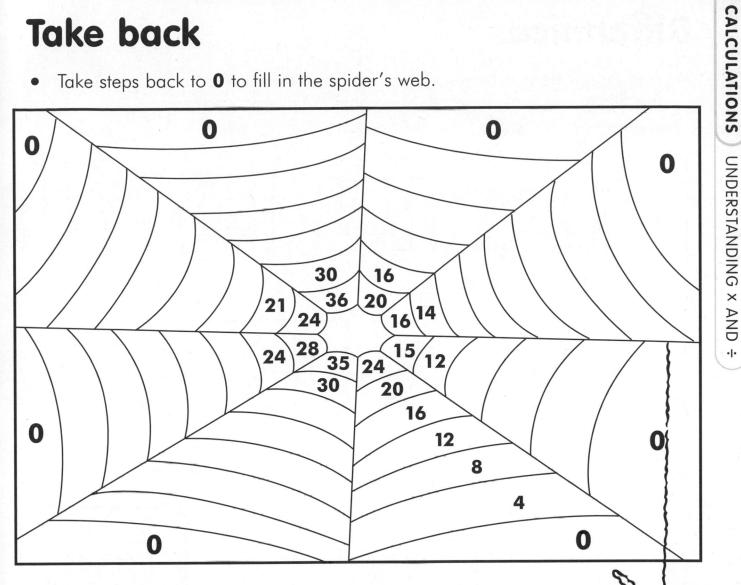

- Use the completed web to help you to write some number sentences like the two shown below:

6 × 4 = 24

24 ÷ 4 = 6

Dear Helper,

This activity provides your child with opportunities to explore further links between multiplication and division. This time division is being explored through 'repeated subtraction' – taking steps back.

Differences

- Find the differences between the numbers at the corners and use this to begin a sequence. Keep going until you reach zero:

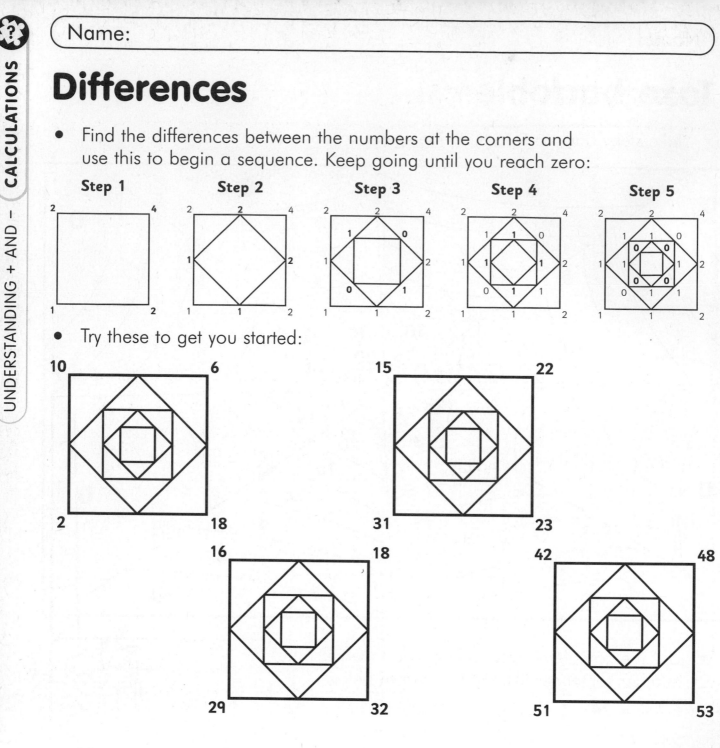

Step 1 **Step 2** **Step 3** **Step 4** **Step 5**

- Try these to get you started:

10   6

2   18

15   22

31   23

16   18

29   32

42   48

51   53

- Now create two of your own:

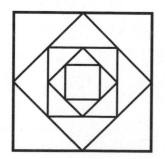

Dear Helper,

This puzzle has been explored in lesson time and has been carried over as homework. The aim in each case is to continue the pattern of 'shrinking squares' until all the differences are 0. This task gives your child lots of practice in subtraction in a fun and interesting way.

Money problems!

I bought:

65p

84p

I spent: _____

Change from £2 _____

Cinema tickets cost £4.50 for adults. Children's tickets are half price.

- How much does it cost altogether for two adults and three children?

I have four coins in my hand. The total is £2.26.

- Draw the coins I am holding.

When I went to the shop with my friend, we spent 70p altogether. If I spent 30p more than my friend, how much did we each spend?

If I save 55p a week for 12 weeks, how much will I save altogether?

Dear Helper,

Solving 'real' problems is important when it comes to spending money. In these exercises, your child may need help in working out what type of mathematical operation to use and how to go about it. In terms of the method, there are no hard and fast rules. One question lends itself to giving change through the 'shopkeeper's method', involving counting on from the total spent to the amount given for payment. Others would benefit from using real coins or from talking things through with others.

Name:

This way, that way

- Find as many different ways of ordering these two sets numbers into sums as you can; and find the total in each case.

13 16 17   **26 36 44**

e.g. **16 + 13 + 17 = 46**

- Now do the same using any three numbers each time. See which sum seems the easiest when you calculate from left to right.

Dear Helper,

When adding together three or more numbers, it is useful for children to know that the order of adding is not important. Indeed, when adding a list of numbers together it is often better to pair numbers together out of sequence for instance, where the total may be 10. In this exercise the numbers are added together in different ways, and then the 'best' order is selected.

Number machines

- These numbers have been partitioned into two numbers and recombined.
- The first one has been done for you. Now fill in the missing numbers on the other machines.

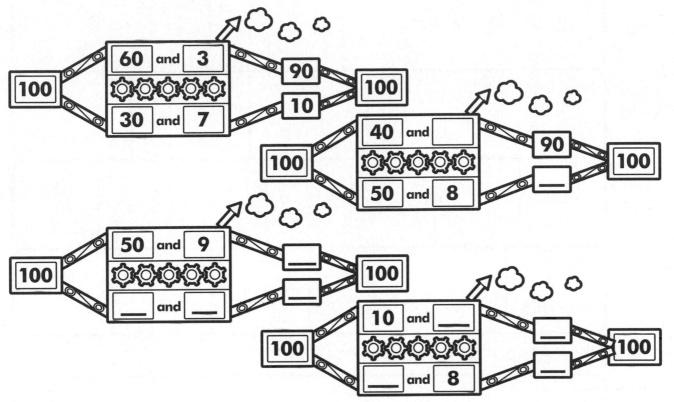

- Create some 'number splits' of your own.

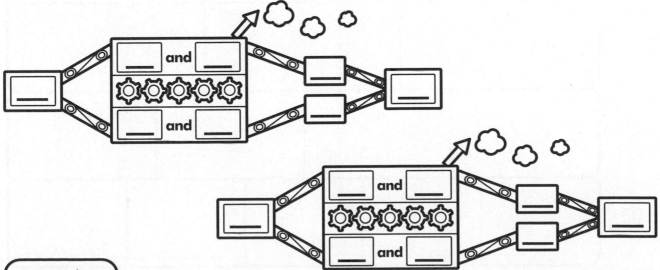

Dear Helper,

It is sometimes helpful to break a number down into smaller units to help with our calculations. A lot of attention is given in lesson time to the idea of splitting numbers into units, tens, hundreds and beyond. This exercise deals with breaking numbers into tens and units and then combining these parts to find the total of two numbers. If your child manages this task with ease, you might encourage them to try to work in a similar way with three-digit numbers.

Name:

Coin combos

- Use £2, £1, 50p, 20p, 5p, 2p and 1p coins to make the prices shown.
- In each case, use just one coin in each box.

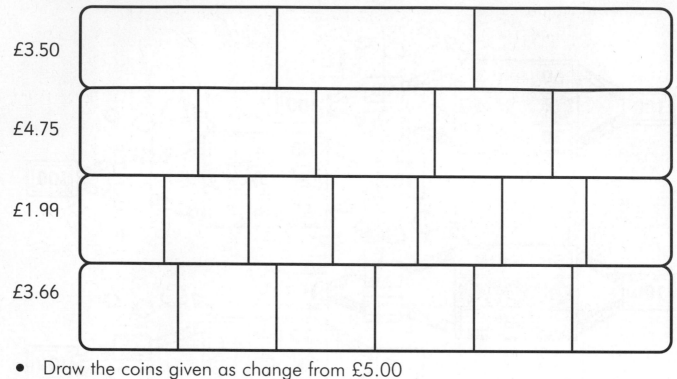

£3.50

£4.75

£1.99

£3.66

- Draw the coins given as change from £5.00

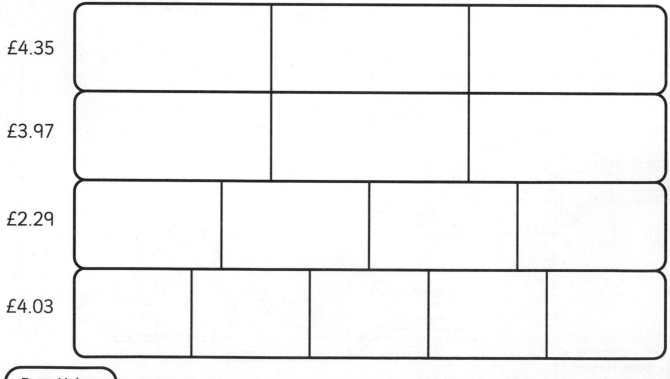

£4.35

£3.97

£2.29

£4.03

Dear Helper,

While we all have to use the change that we have available, it is usually convenient to try to use the least possible number of coins. This task samples this skill by giving the minimum number of coins necessary to make the amounts stated. Using real coins could help here, at each stage trying to use the largest possible coin value without going over the target amount.

Fraction machine

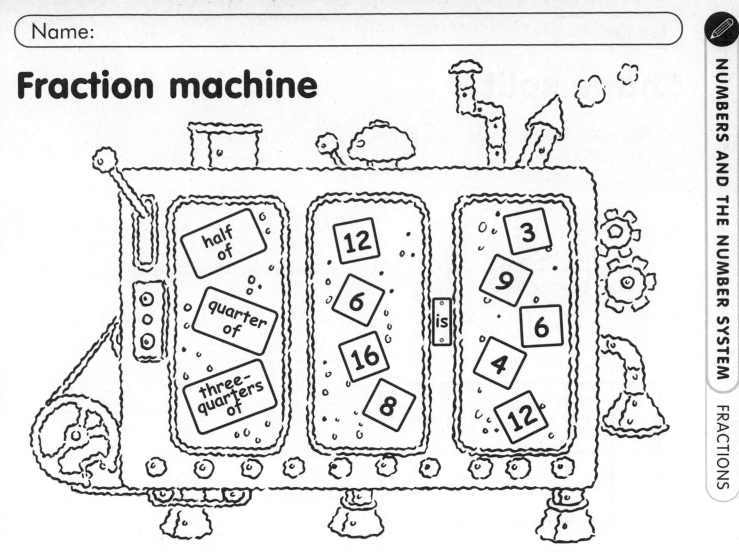

- Make some correct stories from the fraction machine:

e.g. **half of 12 is 6**

Dear Helper,

This activity brings work on fractions, multiplication, division and addition all together since these are all so inter-related. The numbers have been carefully selected so your child should be able to find lots of different statements. Knowing what a quarter of something is can be a useful step to calculating three-quarters of that amount.

Shape split

- For each shape, colour half in red, a
  quarter in yellow and an eighth in blue.

Half of 16 is _____

A quarter of 16 is _____

An eighth of 16 is _____

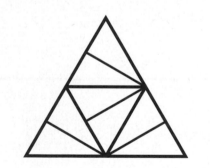

Half of 24 is _____

A quarter of 24 is _____

An eighth of 24 is _____

Half of 8 is _____

A quarter of 8 is _____

An eighth of 8 is _____

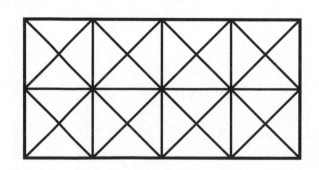

Half of 32 is _____

A quarter of 32 is _____

An eighth of 32 is _____

Dear Helper,

Fractional parts can be found in different ways. A quarter of 12, for example, could be seen as 12 ÷ 4 or
calculated by finding 'half of a half'. Similarly, an eighth could be calculated by 'half of a half of a half'!
Links can also be made between fractions which look different, but actually have the same value: 16 parts
out of 32, for example, is equivalent to ½.

All in a day

- Use one colour to show when you sleep, a second colour to show when you are awake at home and a third colour to show when you are at school.

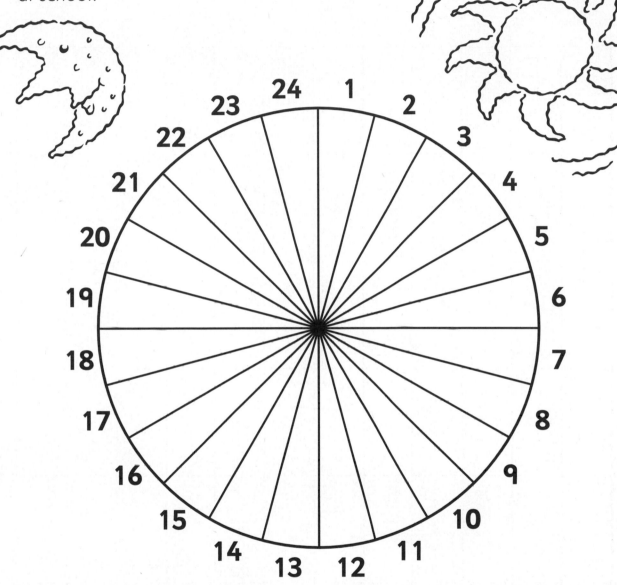

- How many hours are you in bed?_____

- How many hours are you at school?_____

- How many hours does that leave in a day?_____

Dear Helper,

This work should produce a record of a typical day (and night) in a way that could be described as a 'pie chart'. The pie chart shows in a very visual way the proportion of time that we actually spend in bed, something that we are not usually aware of when we are sleeping!

Name:

Birthday chart

- Make a graph of birthday months using the information gathered on your class.

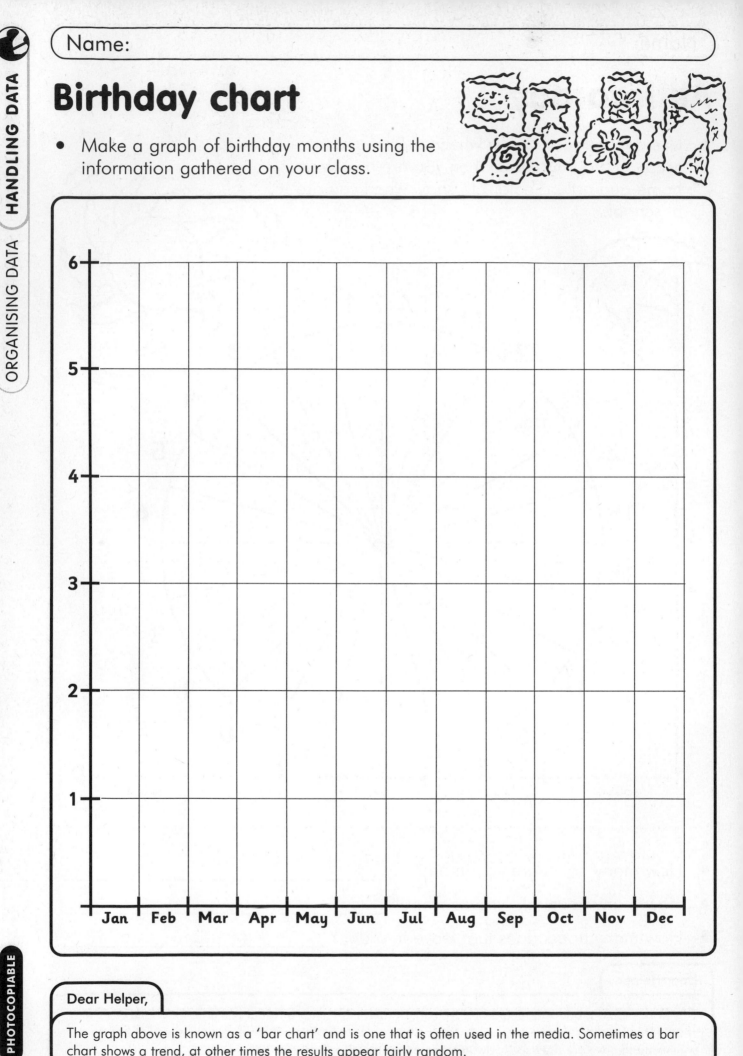

Dear Helper,

The graph above is known as a 'bar chart' and is one that is often used in the media. Sometimes a bar chart shows a trend, at other times the results appear fairly random.

Name:

Pets corner

- Make a graph of favourite pets using information gathered by your class.

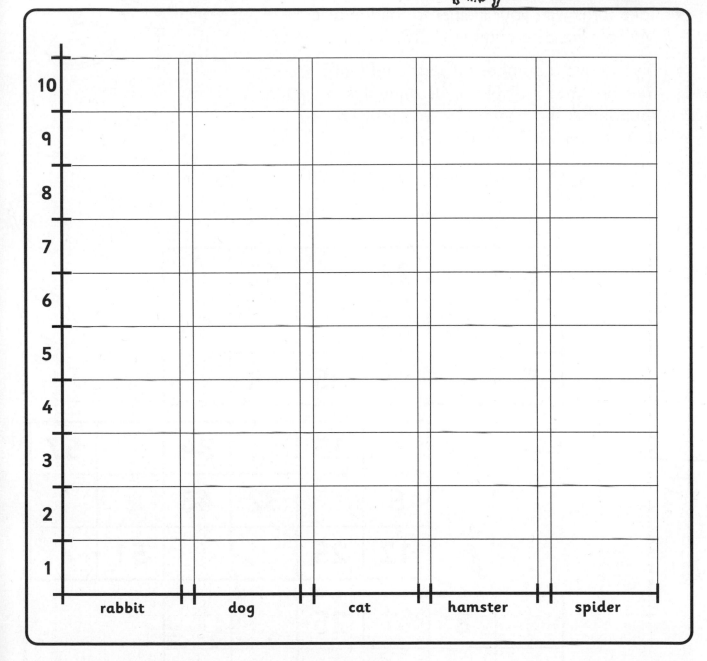

- Write a few statements about the results.

_____

_____

Dear Helper,

This type of presentation is known as a 'block graph'. In this case, each unit block represents a pet. As you would expect, we only show data of this type in whole units – nobody wants to own half of a pet!

Name:

House!

You will need: a game board each, about 30 counters and one standard dice.

- Take turns with your Helper to roll the dice. Multiply the dice number by 10.

- If you have a number on your card that matches your multiple of 10 when it is rounded to the nearest 10, you may cover that answer with a counter.

- The first player to cover all the numbers on their card wins.

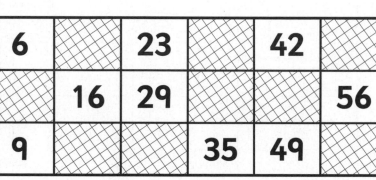

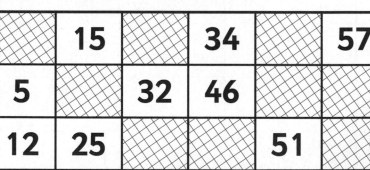

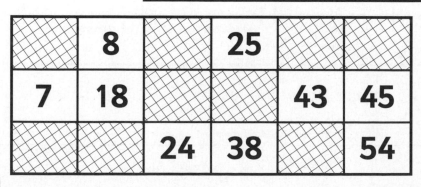

Dear Helper,

The skill of rounding up and down to the nearest 10 is fixed by clear rules. A number such as 64 is rounded down, as it is nearer to 60 than 70. A number such as 67 is closer to 70 than 60 and is, therefore, rounded up. The question of what to do with 65 needs clarifying. Where a number ends in a 5, that number is always rounded up to the nearest 10. Please ensure that your child follows these rules at all times: less than 5 round down, 5 or above round up.

Rough talk

- Pick some tiles from those below to make some number problems for you to solve. Write them in the box below.

- Try to work them out by rounding up or down and then adjusting your answer if you need to.

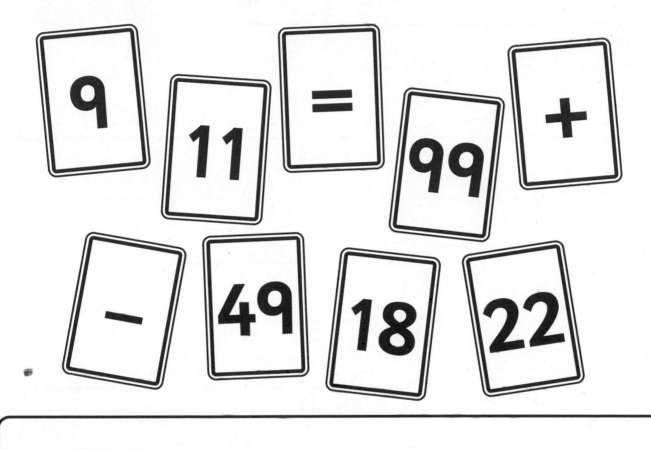

PHOTOCOPIABLE

Dear Helper,

Quite often, we are faced with working out problems of addition and subtraction that do not conveniently involve multiples of 10p or £1.00. Many shop items are priced at levels such as 49p or £4.99– probably to make us feel that we're not spending as much as we are! Having a 'nearly' amount such as 99p is often best thought of as £1.00, and then adjusted down 1p when we've added it to other items. Although this activity isn't about money, the idea is much the same. When adding 18, for example, your child might be encouraged to add 20 in their head and then subtract 2 from the answer.

Staging posts (1)

- Count on from the smallest number to the nearest 10, then on to the finishing number. The first one has been done for you.

$+2$ $+30$ $+4$

18   20   50   **54**   total jump is __**36**__

17   **35**   **23**   **51**

total jump is _____   total jump is _____

33   **76**

total jump is _____

37   **83**   **12**   **91**

total jump is _____   total jump is _____

- Now draw a couple of your own:

total jump is _____   total jump is _____

Dear Helper,

Often, we find the difference between two numbers not by using subtraction, but by counting on from the lower number – the 'shopkeepers method' for giving change. Children make use of number lines in school to make this have more meaning; 'jumping' along the number line and 'stopping off' at convenient points on the journey. The total of all the jumps represents the difference between the two numbers.

Name:

Budget

You have only £10.00 to spend.

- Choose what you would like to buy and check you have not spent too much!

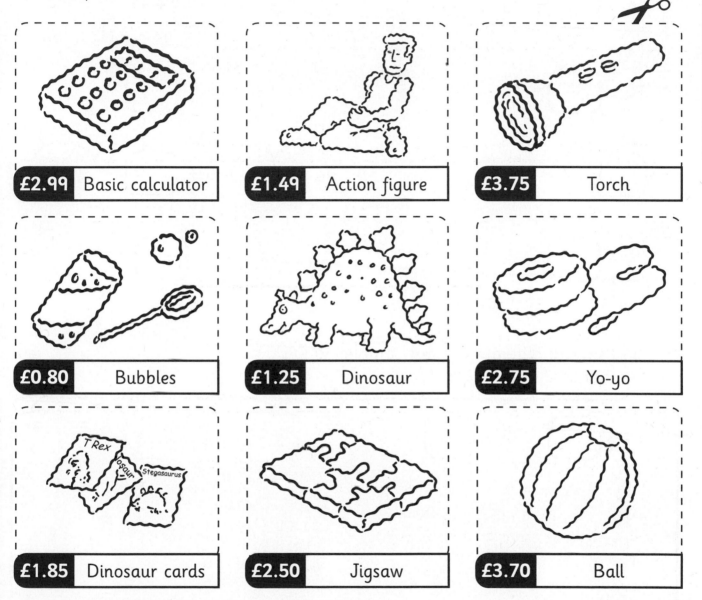

£2.99 Basic calculator	£1.49 Action figure	£3.75 Torch
£0.80 Bubbles	£1.25 Dinosaur	£2.75 Yo-yo
£1.85 Dinosaur cards	£2.50 Jigsaw	£3.70 Ball

- Cut out the items you would like to buy and stick them on to a plain sheet of paper.

- Write down how much you spend and the change left from £10.00.

Dear Helper,

This is a nice problem to have – what to spend your money on! Your child may need some help in keeping the running total as the items accumulate. In this case, you may want to use a calculator (if available), rather like a shopkeeper uses a cash register. It is important that money totals are written in the correct way. Avoid adding 'p' when writing in pounds, so £1.00 and 99p is written as £1.99, not £1.99p. Another mistake is to forget the 0 in amounts just over the pound, for example £4.00 and 5p is written £4.05, not £4.5.

Name:

All things equal

- Find the missing numbers:

14 + ☐ = 20

16 − ☐ = 11

☐ − 6 = 22

24 + ☐ = 39

- Now try these:

16 + 5 = ☐ + 12

13 + 4 = 22 − ☐

29 − 13 = 6 + ☐

7 + 9 = 21 − ☐

- Find several different ways of completing this number sentence:

14 + ☐ = 25 − ☐

Dear Helper,

Work on finding missing numbers is another aspect of early algebra. Activities of this type further reinforce the idea of addition and subtraction being linked. Some children find questions with 'double-ended problems' (either side of the = sign) quite a challenge. Sometimes it helps to think of the = sign as a see-saw, with the two sides being in balance. Most of these questions have just one answer, although you will notice that things get a little more open-ended at the last minute!

PHOTOCOPIABLE

Name:

How's that? (1)

• Answer these questions.

Question	Answer	How I did it
What is 12p less than 29p?		
Subtract 14p from 31p.		
Reduce 49p by 21p.		
How much less is 49p than 81p?		
I spend 16p and I am left with 46p. How much money did I start with?		
How much must be taken from 50p to leave 17p?		
£1.49 less than £5.00 is…?		

Dear Helper,

Sometimes the best way to decide how to work out a mathematics problem is to look at the numbers involved. If we have to subtract 19, for example, it might be best to subtract 20 and then adjust the answer by 1. This exercise aims to encourage your child to get the right answer *and* to explain how they got there.

How much?

These items are sold at various prices:

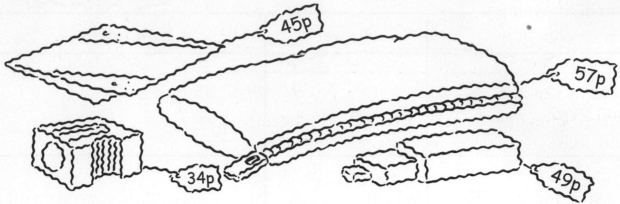

- Use column addition to work out some different totals using any two prices each time. One has been done for you in two different ways:

- Write some more in the box below.

45		45	
+ 49		+ 49	
—		—	
14	and	80	
80		14	
—		—	
94		94	

Dear Helper,

This method of column addition may be different to how you were taught at school. This approach is useful as an early stage before moving on to 'tidy' approaches which allow for 'carrying' when the column totals exceed 9. If you are offering your child support, it will be helpful if you to try to follow one or both of the methods given in the example.

Staging posts (2)

You can calculate 75 – 32 by counting on from
the smaller number:

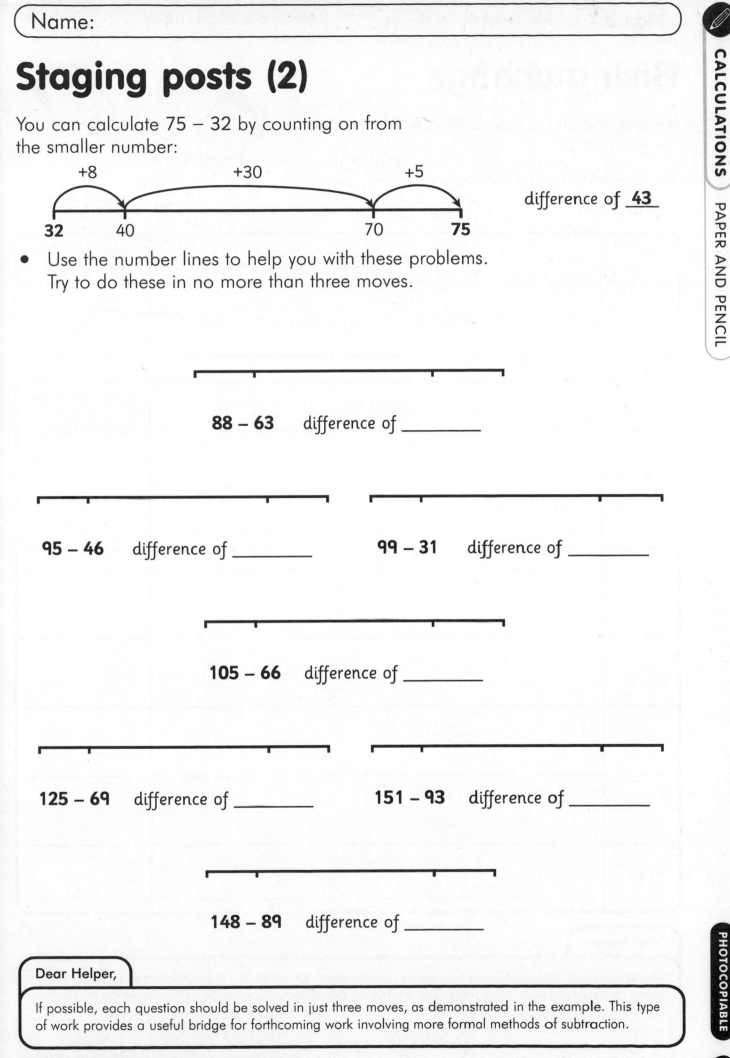

difference of __43__

- Use the number lines to help you with these problems.
  Try to do these in no more than three moves.

88 – 63   difference of _____

95 – 46   difference of _____

99 – 31   difference of _____

105 – 66   difference of _____

125 – 69   difference of _____

151 – 93   difference of _____

148 – 89   difference of _____

Dear Helper,

If possible, each question should be solved in just three moves, as demonstrated in the example. This type
of work provides a useful bridge for forthcoming work involving more formal methods of subtraction.

Giving change

- Use some coins to practise giving change from £1.00.

- Remember to count on from the cost of the item.

Cost of item	Working out	Total change from £1.00
24p	1p and 5p makes 30p, 20p gives 50p, 50p makes £1.00	**76p**

Cost of item	Working out	Total change from £1.00
45p		
67p		
74p		
21p		
39p		
8p		
11p		

Dear Helper,

It is helpful to offer support when your child is working with money, as it is easy for children to forget some of the conventions for giving change. The most important idea continues to be that of using the fewest coins by counting up, using the greatest coin value possible at each stage. The example above demonstrates the preferred method of giving change. If at all possible, let your child use real money.

Name:

Problems, problems!

- Use coins to help you solve these word problems.

How much would three pencils cost at 25p each?	If I spend 85p on an ice-cream, what change should I get from £2.00?
What is the difference in pence between £1.25 and 68p?	Which costs less, two pens at 46p each or three rulers at 29p each?
Three stamps cost 72p. The stamps are all the same value. How much is one stamp worth?	I spend 35p. My friend spends twice as much. How much do we spend altogether?

- Now try to write a money problem for yourself on the back of this sheet. Work out the answer.

Dear Helper,

This homework practises further your child's ability to solve 'real' problems in money. As always, it will be helpful if you can provide actual coins wherever possible. Some children may have difficulty in deciding what number operation (+, −, x, ÷) is required, so it may be necessary for you to talk to your child about the types of words which help us with this. For instance, 'how much more?' often involves counting on or subtraction.

Name:

Solid shapes

- Draw some packages or containers of different shapes in the table below.
- Count how many faces, vertices and edges each one has.

Item	Faces	Vertices	Edges

Dear Helper,

This homework uses the technical language of edges, faces and vertices. The surfaces are known as 'faces', the corners are 'vertices' and the edges are simply 'edges'! Using the home environment to identify solid shapes is a valuable way of making the topic meaningful. It is useful for children to consider why certain packages are the shapes they are. In the case of food products, this is sometimes for convenience of stacking. In other cases, it makes the product more appealing or secure.

Name:

2-D shapes

| circle | triangle | square | rectangle |
| rhombus | parallelogram | trapezium | kite |

- Cut out the shape names above and stick them in the correct spaces under the shape pictures.

Dear Helper,

Your child should be familiar with the names of the most common shapes. These shapes are 2-D and are sometimes referred to as 'plane shapes' since they have no thickness (in theory, at least!). In describing these shapes, we refer to their sides and corners. Some shapes, such as a square, have right-angled corners. Some shapes have pairs of parallel sides, for example the rhombus. Some shapes are inter-related. Did you know, for example, that a square is actually a very special member of the rectangle family?

In balance

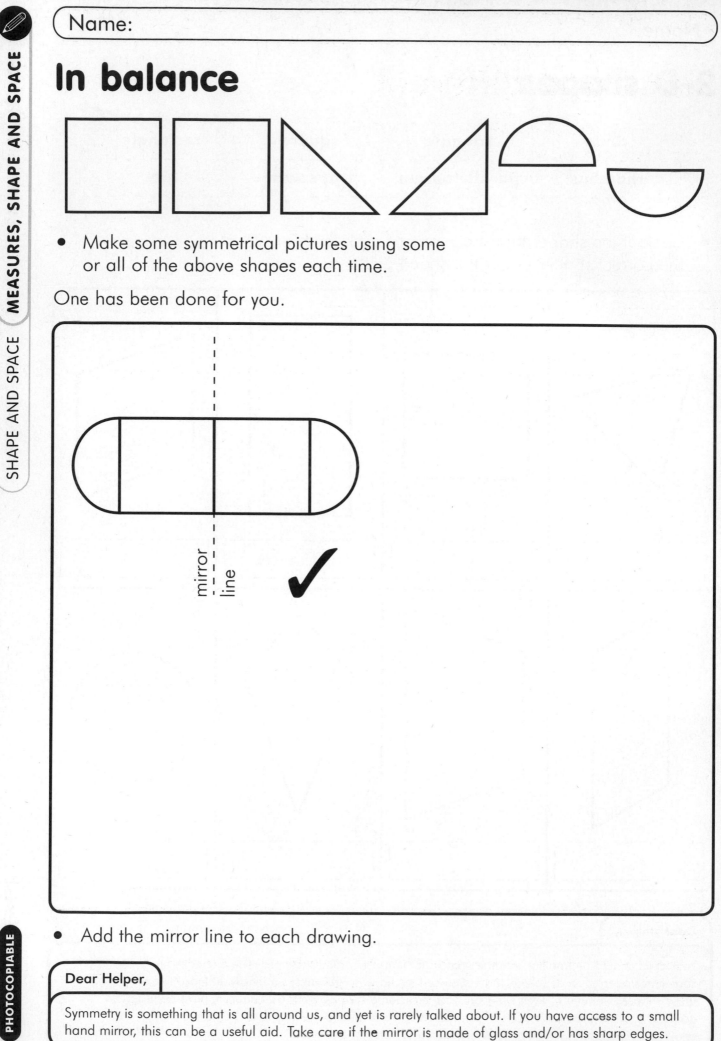

- Make some symmetrical pictures using some or all of the above shapes each time.

One has been done for you.

mirror line ✔

- Add the mirror line to each drawing.

Dear Helper,

Symmetry is something that is all around us, and yet is rarely talked about. If you have access to a small hand mirror, this can be a useful aid. Take care if the mirror is made of glass and/or has sharp edges.

Name:

Same and different

These shapes can be made
using three squares, with
each square being joined
along at least one side:

- Cut out the four squares provided.
- How many different arrangements of them can you find?

Dear Helper,

This activity calls on good visual skills. Sometimes an arrangement looks different when actually it is simply a rotation of one created earlier. On other occasions a new arrangement can be created by forming a mirror image of another. Working in a thorough manner will help children to find all the possible arrangements.

100 MATHS HOMEWORK ACTIVITIES • YEAR 3 TERM 3

All right now

- Cut out this angle measurer and use it to find items around the house with corners that are more, less or the same as a right angle. One has been done for you.

Less than a right angle	A right angle	More than a right angle

Dear Helper,

'Angle' is another form of measurement which is taught as part of mathematics. The idea of 'angle' as a measure is quite difficult for some children to grasp. The angle at any given point has nothing to do with the size of the object: it is purely a measurement of 'turn'. Children are introduced to the right angle with an angular measurement of 90 degrees (90°). This is a common feature in the construction of buildings and objects around our homes.

Name:

Seeing double

- Find some small items around the home and make some measurements of length.

- Use these results to create a double-sized picture of each item.

Dear Helper,

This activity requires accurate measurement of objects using a ruler marked in centimetres and millimetres. By doubling each dimension a double-sized sketch can be created. It is useful to discuss with your child how many times a simple postage stamp would fit inside its double-sized brother – the answer is not twice!

Name:

Treasure island

- Use some of these features to fill some of the squares with pictures on your island map.

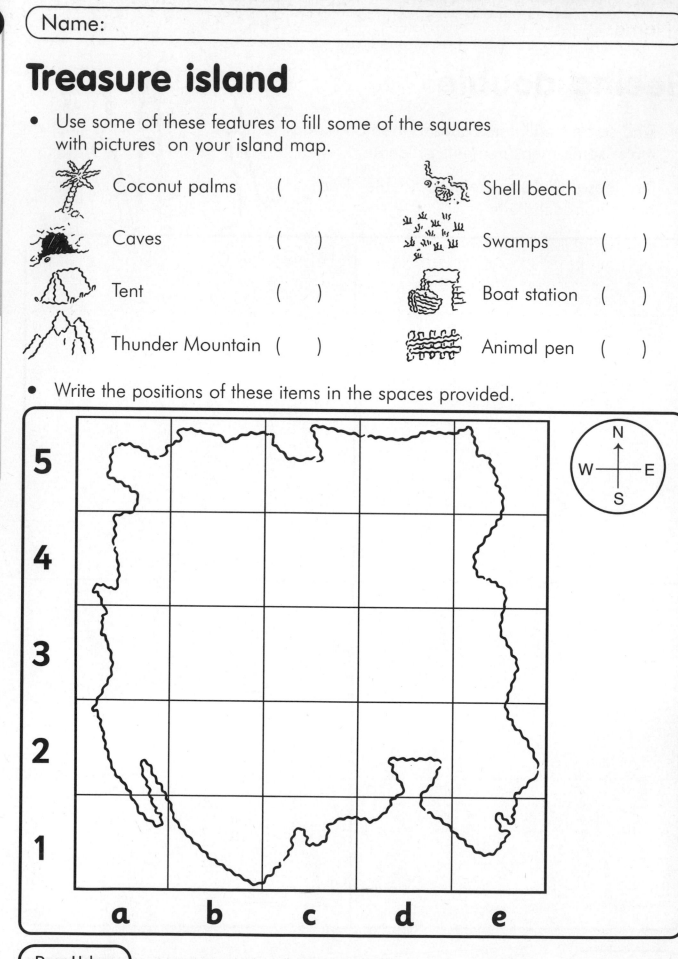

Coconut palms   ( )   Shell beach ( )

Caves   ( )   Swamps ( )

Tent   ( )   Boat station ( )

Thunder Mountain ( )   Animal pen ( )

- Write the positions of these items in the spaces provided.

Dear Helper,

This fun activity requires your child to use their imagination and practises location-finding using grid coordinates to denote the squares. It is useful to draw your child's attention to the four compass directions; something that will be developed further in school.

Name:

In a spin

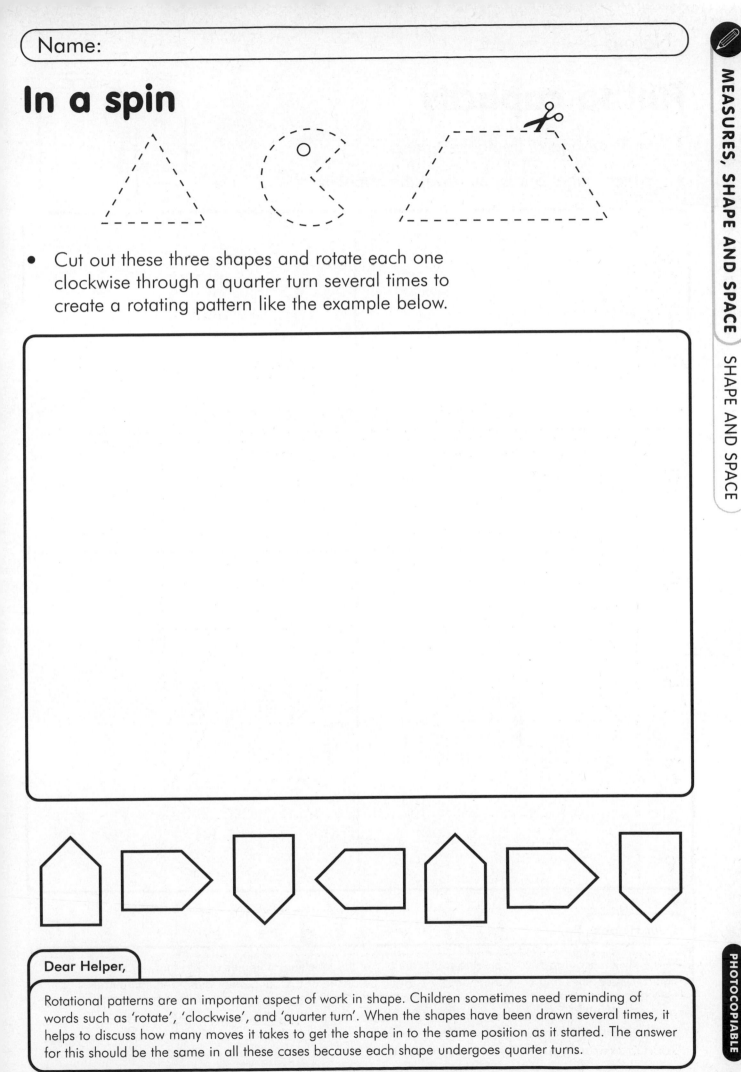

- Cut out these three shapes and rotate each one clockwise through a quarter turn several times to create a rotating pattern like the example below.

Dear Helper,

Rotational patterns are an important aspect of work in shape. Children sometimes need reminding of words such as 'rotate', 'clockwise', and 'quarter turn'. When the shapes have been drawn several times, it helps to discuss how many moves it takes to get the shape in to the same position as it started. The answer for this should be the same in all these cases because each shape undergoes quarter turns.

Full to capacity

- Find a range of liquid containers.
- Sketch, label and write down the capacity of each one.

Dear Helper,

This activity encourages your child to look carefully at the labelling of liquids and to find out how much liquid these bottles and containers hold. As usual, be aware of the dangers of glass and some household liquids such as bleach. There are three main units of capacity found on household containers. The smallest measure, the millilitre (ml), is so small that 1ml would only partially fill a teaspoon! The centilitre (cl) is a less familiar unit for many people and is equivalent to 10ml. One litre (l) is equivalent to 1000ml or 100cl. Some family-sized fizzy drink bottles contain 1, 1.5, 2 or even 3 litres of liquid.

Name:

Number sort

| 2 | 5 | 7 | 15 | 20 | 24 | 40 | 55 | 29 |

- Cut out and paste these numbers in the correct places on the diagram.

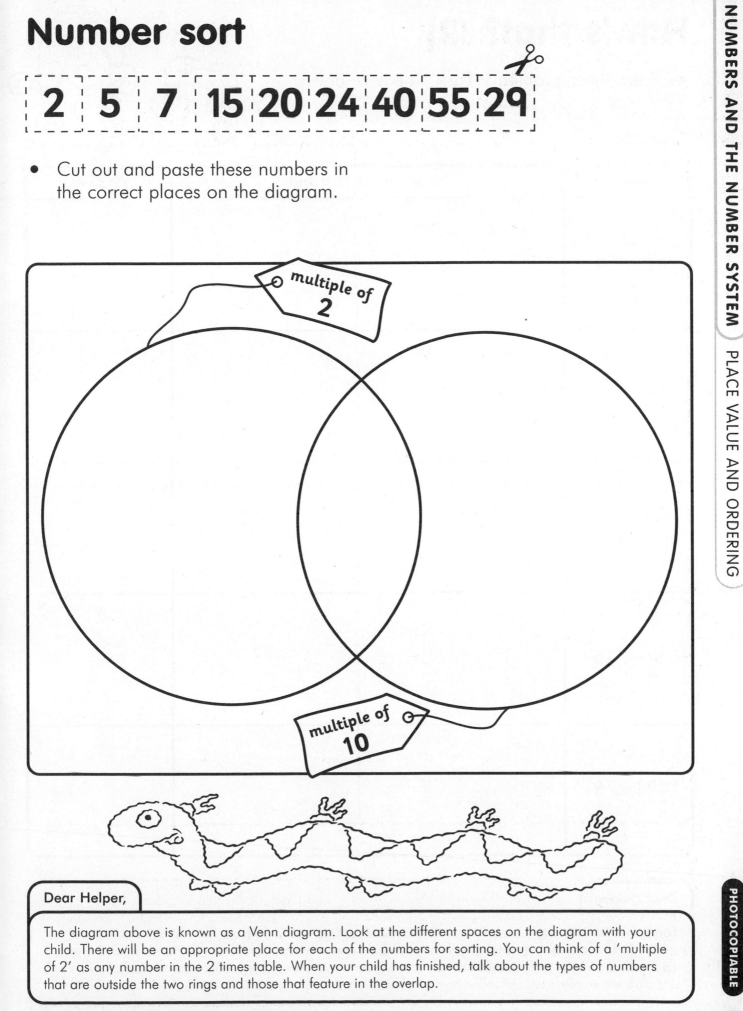

multiple of
2

multiple of
10

Dear Helper,

The diagram above is known as a Venn diagram. Look at the different spaces on the diagram with your child. There will be an appropriate place for each of the numbers for sorting. You can think of a 'multiple of 2' as any number in the 2 times table. When your child has finished, talk about the types of numbers that are outside the two rings and those that feature in the overlap.

Name:

How's that? (2)

- Solve these problems. Each time write down, in words, how you got your answer.

Question	How I worked it out	Answer
Half of 48		
54 times 2		
Share 64 by 4		
8 times 15		
119 less 51		

Dear Helper,

Sometimes the best way to decide how to work out a mathematics problem is to look at the numbers involved. If we have to multiply by 4, for example, it might be best to simply double the number and then double it again. This exercise aims to encourage children not only to get the right answer, but also to explain their strategy.

Odds and evens

- Make up some addition problems of your own and work out the answers.

odd number + odd number	even number + even number	even number + odd number

- Look at your answers in each column. What do you notice?

Dear Helper,

In order to notice a pattern, it may be necessary to try lots of different examples. Look at the numbers in any one column. Your child needs to consider if they have anything in common. This is designed to introduce the idea of a 'general statement', (i.e., one that works for all cases of the same type).

Name:

Starting grid

- Can you place the numbers provided into the grid so that each column and the bottom row always come to the same total?

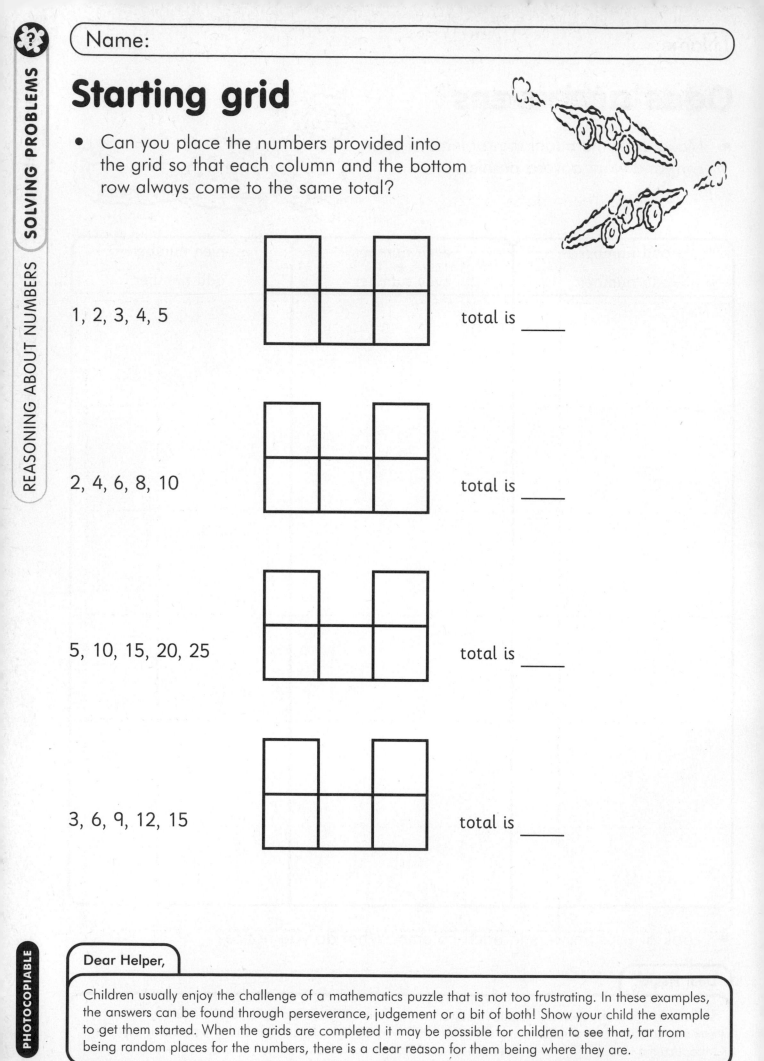

1, 2, 3, 4, 5   total is ____

2, 4, 6, 8, 10   total is ____

5, 10, 15, 20, 25   total is ____

3, 6, 9, 12, 15   total is ____

Dear Helper,

Children usually enjoy the challenge of a mathematics puzzle that is not too frustrating. In these examples, the answers can be found through perseverance, judgement or a bit of both! Show your child the example to get them started. When the grids are completed it may be possible for children to see that, far from being random places for the numbers, there is a clear reason for them being where they are.

Name:

What's the story?

- Can you turn each of these number sentences into a word problem?

The first one has been done for you.

		Answer
36 – 25	I had 36 stickers, but I have used 25. How many are left?	11
41 + 39		
54 – __ = 32		
__ + 46 = 98		
6 x 5 = __		
42 ÷ 2 = __		
14 + 67 – 44 = __		

Dear Helper,

This task is designed to see if the children can 'wrap' a pure mathematics question in a 'real' context. While your child is creating these number stories, encourage their ideas involving money, length, time and other measures.

Name:

What's it worth?

- Imagine you have been given three coins of the same value.

- Write down all of the different amounts that you could be holding. One possible amount has been done for you.

	Amount in pence	Amount in pounds
2p and 2p and 2p	6p	£0.06

Dear Helper,

Your child is not expected to find all the possible combinations for this activity as there would be insufficient space (and probably time) to record them all. As is often the case in money questions, it is helpful to have real coins on hand. The key thing about this puzzle is to encourage your child to record their findings using both the money conventions detailed in the table: 'p' and '£0.00'.

Maths links

- Write four number sentences for each family of numbers.

The first one has been done for you.

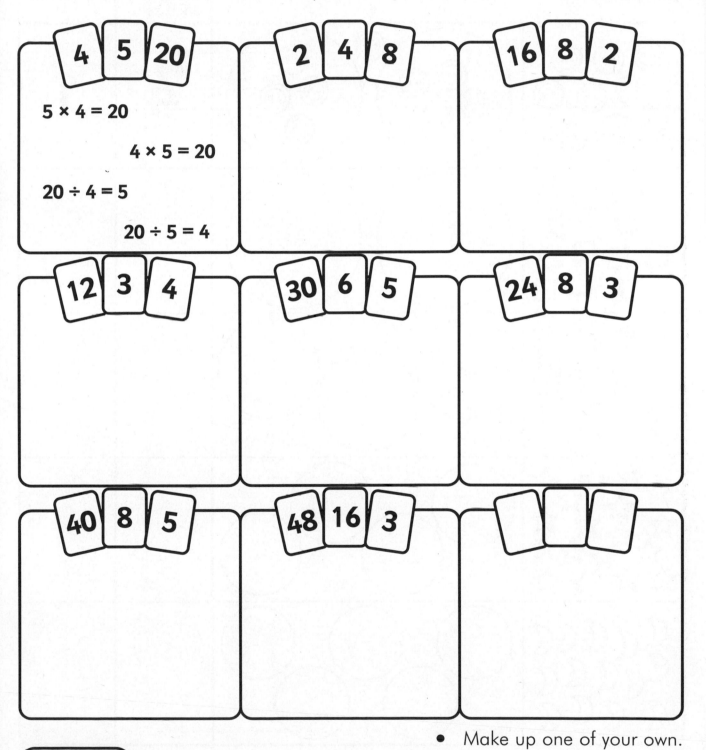

4 5 20

$5 × 4 = 20$

$4 × 5 = 20$

$20 ÷ 4 = 5$

$20 ÷ 5 = 4$

2 4 8

16 8 2

12 3 4

30 6 5

24 8 3

40 8 5

48 16 3

- Make up one of your own.

Dear Helper,

In just the same way as addition and subtraction are linked, so too are multiplication and division. This exercise gives further practice in establishing that link. Sometimes it helps to talk about what these multiplications and divisions might mean in a real situation, otherwise creating these number arrangements can just become a trick with little real understanding.

Name:

Remainders (1)

- Share these objects into equal groups and find the remainder.
  The first one has been done for you.

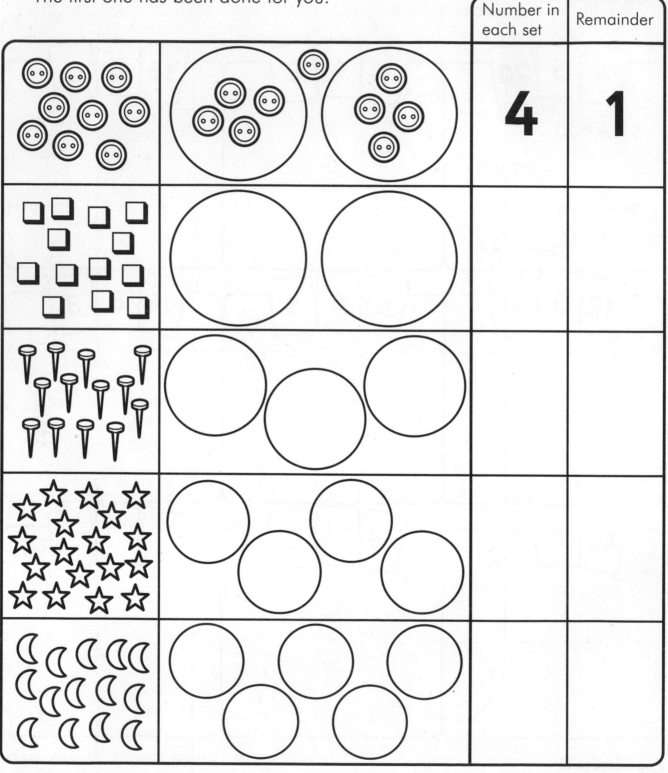

		Number in each set	Remainder
		4	**1**

Dear Helper,

Your child will benefit from sharing real objects. If small counting objects such as counters, buttons or coins are available these should be used. This type of work helps to support related skills such as times tables work, and also prepares your child for more formal work on division.

Name:

Remainders (2)

- Use the information in the last two columns to draw objects in the second column, and then in the first column.
  One has already been done for you.

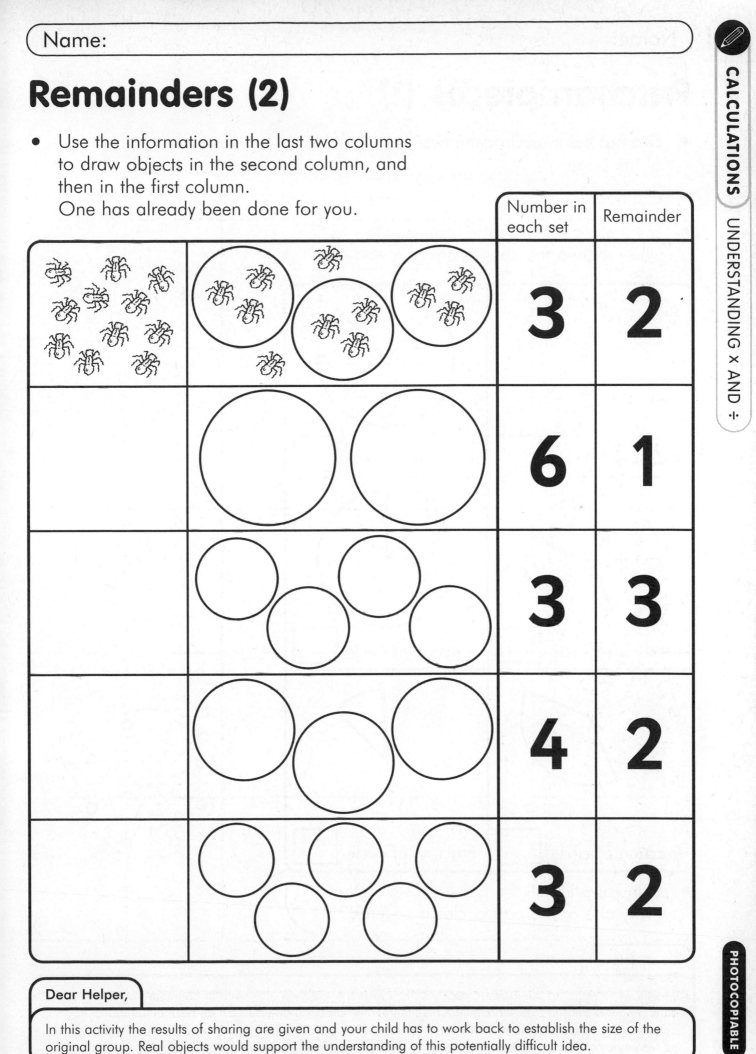

		Number in each set	Remainder
		3	2
		6	1
		3	3
		4	2
		3	2

Dear Helper,

In this activity the results of sharing are given and your child has to work back to establish the size of the original group. Real objects would support the understanding of this potentially difficult idea.

Name:

Fraction pieces (1)

- Cut out the three fraction wheels at the side of the page.

- Cut each one up into its individual parts.

- Lay these parts onto the shapes below to see what fraction the shapes are of a whole.

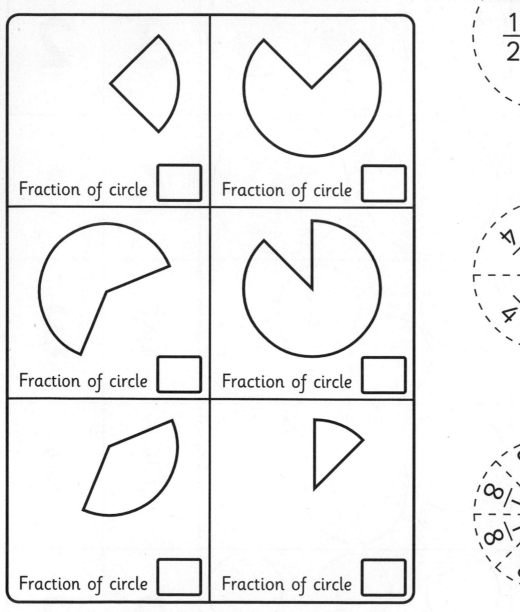

Fraction of circle ☐

Fraction of circle ☐

Fraction of circle ☐

Fraction of circle ☐

Fraction of circle ☐

Fraction of circle ☐

- Can you stick all the pieces down so that every one of the part circles is covered?

Dear Helper,

As well as recognising the familiar fractions of ½, ¼ and ⅛, children also need to combine fractions with different denominators. The denominator is the bottom number of the fraction and is the number of equal parts that the shape is divided into. In ⅛ the denominator is 8. You can support your child by ensuring that all the circles can be completed before sticking the pieces down permanently.

Name:

Which is bigger?

- Use the fraction wall to make some 'greater than' statements.

  One has been done for you.

Fraction wall

one whole							
$\frac{1}{2}$				$\frac{1}{2}$			
$\frac{1}{4}$		$\frac{1}{4}$		$\frac{1}{4}$		$\frac{1}{4}$	
$\frac{1}{8}$	$\frac{1}{8}$	$\frac{1}{8}$	$\frac{1}{8}$	$\frac{1}{8}$	$\frac{1}{8}$	$\frac{1}{8}$	$\frac{1}{8}$

$\frac{3}{8} > \frac{1}{4}$

Dear Helper,

The fraction wall illustration (above) provides a useful way of comparing fractions. Children need to recognise that the more the sub-divisions, the smaller the pieces become. The numerator tells us how many of the parts of the whole are involved, so in ⅜ the numerator is 3. This activity encourages your child to work with numerators other than 1. The symbol > means 'more than' or 'greater than'.

Name:

Fraction pieces (2)

- Cut each piece from the wall to find out what
  fraction of the rectangles below are shaded.

Fraction shaded _____

Fraction shaded _____

Fraction shaded _____

Fraction shaded _____

Fraction shaded _____

Fraction shaded _____

- Can you stick all the pieces
  down so that all the shaded
  areas are covered?

Fraction wall

$\frac{1}{2}$		$\frac{1}{2}$	
$\frac{1}{4}$	$\frac{1}{4}$	$\frac{1}{4}$	$\frac{1}{4}$
$\frac{1}{8}$ $\frac{1}{8}$	$\frac{1}{8}$ $\frac{1}{8}$	$\frac{1}{8}$ $\frac{1}{8}$	$\frac{1}{8}$ $\frac{1}{8}$

Dear Helper,

This activity involves cutting and sticking. You can support your child by ensuring that all the fractions can
be completed before they stick the pieces down permanently.

Name:

Fraction match

- Cut out the tiles below and group them into three families of equivalent fractions.

- Now paste them on to another sheet of paper.

$\dfrac{3}{4}$	$\dfrac{1}{2}$	half
0.5	$\dfrac{1}{4}$	$\dfrac{2}{4}$
$\dfrac{6}{8}$	0.75	three - quarters
$\dfrac{2}{8}$	quarter	0.25

Dear Helper,

Each of the three fractions is written in three different ways:
- as a conventional fraction
- in words
- as a decimal fraction

Making links between different representations of the same thing is important – fractions can be presented in all of these ways.

Number crunch

- Add any three numbers together in different ways.
- Explain which order was the easiest to calculate.

Numbers	Order 1	Order 2	Order 3	Explanation
14 12 16	14 + 16 + 12 = 42	12 + 16 + 14 = 42	16 + 12 + 14 = 42	The first one was easy because 4 and 6 makes 10 so 14 + 16 makes 30.

Dear Helper,

Addition does not have to be done in the order in which it is written. This activity builds on earlier work undertaken by your child, although it requires more independence in selecting a range of numbers to add together.

Name:

Take that

- You can calculate **125 – 47** by counting on from the smaller number:

$$
\begin{array}{r}
125 \\
-\ 47 \\
\hline
\end{array}
$$

3 (to 50)
70 (to 120)
5 (to 125)
_____
78

- Now try these for yourself:

130 – 89	172 – 54
145 – 60	200 – 77
136 – 49	221 – 137

- Work out a few of your own on the back of the sheet.

Dear Helper,

This exercise practises the method of subtraction favoured for this age group. Following through the example, you will notice that this approach is closely linked to earlier ideas of counting on. This prepares the ground for more formal approaches to subtraction which are encountered later.

PHOTOCOPIABLE

125

Time diary

- Create a picture diary of six different things you will do this weekend, showing the time when each one will happen.

- Show the time on the clock face and in words. For example: *20 minutes past 6.*

Event	Time	Time in words

Dear Helper,

Even though we are in the 21st century, the traditional clock face is still a popular way of telling the time. For that reason, it is essential that both analogue (clock-face) and digital time are understood.

100 MATHS HOMEWORK ACTIVITIES • YEAR 3 TERM 3

Name:

Drinks all round

- Collect data on favourite drinks involving up to five people.

- For each person, enter 1st, 2nd, 3rd and 4th in the columns to indicate their favourite, through to their least favourite drink.

Name	Orange juice	Tea	Coffee	Milk

- Which drink is the most popular?

Dear Helper,

This activity needs the support of family and friends. The selection should be made from the options provided, and not extended to include other favoured drinks. Your child may need help in making sense of the data. Get them to think particularly about how to answer the last question in a mathematical way.

Name:

Favourites

- Carry out a simple survey of favourite things such as colours or cars.

- Write the headings along the top of the chart and collect the information from family and friends.

Name					

Dear Helper,

This final task places a lot of responsibility on your child, so it would be helpful to talk through the sorts of surveys to which people can readily respond. When collecting data of any sort on individuals, it is essential that the data is not of a sensitive or confidential nature.